Environmental Science and Engineering

The ultimate goal of this series is to contribute to the protection of our environment, which calls for both profound research and the ongoing development of solutions and measurements by experts in the field. Accordingly, the series promotes not only a deeper understanding of environmental processes and the evaluation of management strategies, but also design and technology aimed at improving environmental quality. Books focusing on the former are published in the subseries Environmental Science, those focusing on the latter in the subseries Environmental Engineering.

Zhanhong Wan
Editor

Water Resources Management and Water Pollution Control

Conference Proceeding of 2023 the 6th International Symposium on Water Pollution and Treatment (ISWPT 2023)

Springer

Editor
Zhanhong Wan
Institute of Naval Architecture and Ocean Engineering
Ocean College
Zhoushan, China

ISSN 1863-5520 ISSN 1863-5539 (electronic)
Environmental Science and Engineering
ISBN 978-3-031-53455-3 ISBN 978-3-031-53456-0 (eBook)
https://doi.org/10.1007/978-3-031-53456-0

This Springer imprint is published by the registered company Springer Nature Switzerland AG
The registered company address is: Gewerbestrasse 11, 6330 Cham, Switzerland

Paper in this product is recyclable.

Preface

This book encompasses the carefully chosen peer-reviewed proceedings from the 6th International Symposium on Water Pollution and Treatment (ISWPT 2023).

Featuring contributions from researchers, practitioners, policymakers, and entrepreneurs, the book delves into recent advancements in water pollution and treatment. The covered topics include the intersection of water pollution and climate change (technologies for reducing greenhouse emissions in water and wastewater treatment), water resources planning and management, water quality and protection, as well as technologies and processes controlling water pollution and solid waste management.

Designed to cater to beginners, researchers, and professionals engaged in the field of water pollution management, policy, and governance, this book offers valuable insights and comprehensive information.

Dr. Zhanhong Wan
Ocean College
Zhejiang University
Zhoushan, China

Contents

Chapter 1
Assessing Surface Water Vulnerability Zones in Mahanadi River Basin, Odisha, India: An Integrated Approach Using GIS and MCDM

Abhijeet Das and Milad Khatib

Abstract Evaluation of surface water and their controlling processes are very important for the sustainable utilization of surface water in any region. Analytical techniques coupled with GIS, entropy-based water quality index (EWQI) and Multicriteria decision making methods (MCDMs) have been used to identify suitable surface water sites in Mahanadi River, Odisha for drinking purposes. A total of 20 samples from 19 stations were collected and analyzed for 8-year (2015–2023). The study revealed that in the investigated area, 84.20% of the final EWQI value can be classified as having excellent potential, while 10.53% are categorized as having poor potential. Moreover, 5.26% are identified as potential for unsuitable for drinking. COPRAS and GRA analysis is used to validate these findings, demonstrating that both these techniques, achieves an accuracy of 94% in identifying surface water potential zones in the region. This study utilizes the best method derived from both models to identify 19 suitable locations for best drinking water sites. The overall results suggest water quality in most parts of the investigated region is suitable for human consumption, with an exception of ST-(8), (9) and (19). According to the research, anthropogenic and geological activity have negatively impacted these sites, and a proper management strategy is required to prevent additional contamination and pollution of the local surface water.

Keywords GIS · EWQI · MCDMs · Mahanadi River

A. Das
Department of Civil Engineering, C.V. Raman Global University (CGU), Bhubaneswar, Odisha 752054, India

M. Khatib (✉)
School of Engineering, Lebanese International University (LIU), Mouseitbeh, Beirut, Lebanon
e-mail: milad.khatib@liu.edu.lb

Issam Fares Faculty of Technology, University of Balamand, Souq El-Ghareb, Lebanon

Z. Wan (ed.), *Water Resources Management and Water Pollution Control*, Environmental Science and Engineering, https://doi.org/10.1007/978-3-031-53456-0_1

1.1 Introduction

Water occupies 70% of the planet's surface, which is necessary for every process that keeps life alive and is essential to every biological community. Hence, water quality is said to be a state of water system, which exhibited through physical, chemical, and biological parameters (Roy et al. 2022). Due to its potential for less contamination and greater storage capacity than ground water, surface water is regarded as a significant supply of water. Despite the restricted water supply, the need for water is growing daily (Nguyen et al. 2022). Water is being exploited in many parts of the world because to rising human demand for drinking, agricultural, urban development, and industrial uses. However, water quality assessment tries to pinpoint the causes of water contamination and provide a plan for long-term management of water sources. In contrast, a country's ability to grow, depends on how well its water resources are used and conserved (Thacker et al. 2022).

These surface water resources are being misused as a result of the rapid rise in human population, urbanization, and industrialization (Omar et al. 2020). Further, surface water is rapidly being used in many regions of the world, which is causing water levels and quality to drop (Das 2023a). Moreover, hydrochemistry is influenced not just by natural elements like climate, geology, and geography, but also more and more by human inputs (Das 2023b). Therefore, river monitoring is crucial for providing accurate information on their water quality as well as for preventing and controlling pollution (Nyakundi et al. 2022). In addition, the budget, withdrawal rate, and other crucial details about surface water flow conditions are crucial for the development of surface water resources (Das 2023c).

In the recent past, GIS-based water quality index has also been used by a few researchers to assess surface water quality and examine the spatial heterogeneity of surface water quality indicators (Das et al. 2021). Further, GIS based mapping of surface water of chemical parameters are crucial for the sustainable management of surface water resources and serve to identify any region's degraded zones (Das 2023d). Since it is practically difficult to collect field data for every location in the research area, data from sampled sites were utilized to estimate the data for un-sampled points using spatial interpolation techniques. Inverse Distance Weighting (IDW) is a popular technique for spatial interpolation (Zade et al. 2021). IDW is a deterministic method of interpolation that uses the proximity of the observed points to estimate value at unmeasured places (Das 2022a). On the other hand, IDW was used to interpolate the surface water quality parameters that did not follow a near normal distribution (Das 2022b). This method effectively captures the extreme values that are more common when dealing with surface water quality data (Goswami and Ghosal 2022).

In the ongoing study, the water quality of the Mahanadi River in Odisha has been evaluated, and potential causes of contamination have been found to better understand the situation. To provide a simple and valuable tool for decision making on the surface water quality, an integrated approach of both entropy water quality index (E-WQI), multi-criteria decision approach (MCDM) and geographical information system (GIS) can be used. Entropy theory has been utilized frequently over the past ten years to analyze water quality and has been found to be more accurate than other indexing techniques. As a result, this method eliminates human bias and is highly objective in weight calculation (Allafta et al. 2021). Thus, this approach to weight calculation eliminates human subjectivity and is highly objective. However, this technique is a useful method for describing the quality of water by condensing a lot of data about the quality of the water into a single, understandable number that represents a regional index value (Agarwal and Tayal 2023). MCDM approaches have found an extensive range of applications in water quality estimation as well as for household regions in recent years.

Additionally, MCDM makes decision-making simple by determining the impact of uncertainties that frequently characterize problems with water management. In the current inquiry, Complex Proportional Assessment (COPRAS) and Grey Relational Analysis (GRA) have been used to tackle issues with water resources. Similar to this, the GIS technique has successfully demonstrated its utility for mapping surface water quality. Additionally, it can be used to categorize surface water as useful and unwanted for farming and consumption purposes (Qadir et al. 2020). Previous researchers in different studies, already use it to get fantastic results (Kumar et al. 2022). Despite the fact that numerous studies have concentrated on the solid waste of the dumpsite and its effects on the Mahanadi River's water quality, Odisha, hence, there is no research on the variations in water quality caused by a confluence of all conceivable natural and manmade factors (Dibs et al. 2023). These studies were primarily concerned with a single method of solving problems in a given area, which left a research gap that needed to be filled by creating a novel integrated strategy. As a result, this study demonstrates the value of this novel integrated approach (Gupta et al. 2020). Therefore, the purpose of the study was to use Entropy, MCDM, and GIS in conjunction to conduct an investigation into and interpretation of the surface water quality, as well as to delineate the areas where the water is appropriate or unfit for drinking.

1.2 Study Area

In this paper, the Mahanadi River was selected as the study region to assess the water quality. The river basin lies within 80° 30′ E–86° 50′ E and 19° 20′ N–23° 35′ N, that covers approximately an area occupying 141,600 km^2, and an annual discharge of $50 * 10^9$ m^3 (Fig. 1.1).

From the Amarkantak hills of the Bastar plateau in Madhya Pradesh State, it emerges as a little pool. The yearly discharge of this river, which is the greatest in Odisha, is around 66,640 Mm^3. The study river is 851 km long overall, 357 km of which are in the state of Chhattisgarh, and the remaining kilometers are in the state of Odisha (Das 2023d). The average annual rainfall is 1572 mm, of which the southwest monsoon, which lasts from mid-June to mid-October, accounts for 70%. As it runs through the districts of Cuttack and Puri from west to east

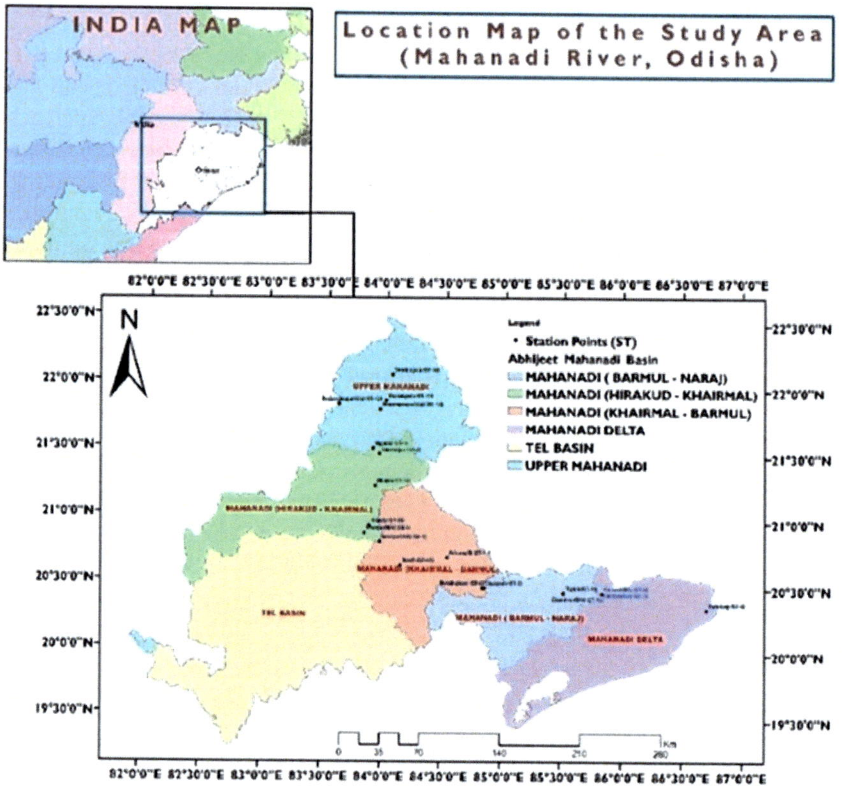

Fig. 1.1 Location map of Mahanadi River, Odisha

through numerous distributaries, it has developed into a sizable delta (Das 2022a). Agricultural runoff and human impacts are evident at ST- (2), (8) and (9), where there has been a significant increase of industry and sewage discharges. The estuary environment of the river was modified by port and human activity (Das 2022b).

1.3 Sampling and Testing

A total of 20 surface water samples were collected from the present study area along 19 different points along the river for 8-year (2015–2023) period. The longitude and latitude of each site were obtained from Google Maps, and the precise positions of each site were noted. The collected samples were stored in an ice box and transported to the laboratory for the analysis of various physical and chemical parameters such as Iron (Fe^{2+}), Fluoride (F^-), Chloride (Cl^-), pH, Total Coliform (TC), Total Hardness (TH), Total Dissolved Solids (TDS), Boron (B^+), Sodium Adsorption Ration (SAR), Total Suspended Solids (TSS), Total Alkalinity (TA), Nitrate (NO_3^-), Dissolved

Oxygen (DO), Chemical Oxygen Demand (COD), Ammoniacal Nitrogen (NH_3-N), Total Kjehdahl Nitrogen (TKN), Electrical Conductivity (EC), Sulphate (SO_4^{2-}), Biochemical Oxygen Demand (BOD), and Free Ammonia (Free-NH_3) using APHA (APHA 2012) standard methods.

Every sample was preserved at 4°C until laboratory analysis. The selection of parameters should depend on various factors such as geographical condition of the study area, a literature review and expert opinions, and the intended use of water. pH and DO were measured at the site itself using pH and DO meter. EC, TSS and TDS were examined, using a water quality analyzer kit in the field, while the Cl^-, SO_4^{2-}, NO_3^- and F^- measured using Ion chromatography; Fe^{2+} and B^+ were calculated, using atomic absorption spectroscopy; alkalinity, Free-NH_3, NH_3-N, TKN and hardness was established, using the titration by H_2SO_4; and SAR was examined using, Flame Emission photometric method. Further, BOD and COD were determined by Winkler's method. To confirm the precision and accuracy of the measurements, QA/QC (quality assurance/ quality control) procedures were made by analyzing blanks and duplicate samples. Water quality parameters were cross-checked, using ionic balance error (IBE) for their precision in analyzing chemical data, where the cations and anions are articulated in milligrams per liter (mg/l). The value of IBE as detected in the current study, is below the allowable level of 5%. Spatial diagrams were used to analyze the results, and they were contrasted with World Health Organization (WHO 2017) drinking water guidelines.

1.4 Methodology

The surface water quality for this investigation was modelled using three effective techniques, i.e., entropy-based water quality index (E-WQI), Complex proportional assessment (COPRAS) and Grey relational analysis (GRA) approach. However, these methods have been discussed in the following sections. The following method, based on Shannon information entropy, is used to calculate E-WQI (Shannon 1948), has been adopted in the ongoing work. This method attempts to offer a better way of providing a numerical expression that is cumulatively derived and describes a certain water quality level based on information entropy (Egbueri 2022). Hence, assessing the spatiotemporal variability of a water source's water quality is crucial. A simple method for conducting these evaluations is to use a single numerical score or index.

For instance, reduce the indicator weights by Shannon's entropy, and more effective information will result, and vice versa (Liu and Ye 2023). This kind of process is suggested to be an object-emancipating technique. It can be said that one may argue that the degree of uncertainty reduction can be used to indirectly study the quantity of information; that is, a reduction in uncertainty that is greater than entropy tends to be more spatially resolved (Ji et al. 2023). On the other hand, this calls for the integration of these sizable and intricate data sets into useful findings that can both inform planners and decision-makers about the general state of a water body's water quality and enable them to take corrective action in the event of contamination

(Chang et al. 2023). The following equations are included in the creation of E-WQI, as recommended by Singh et al. (2020). Finally, Finally, the E-WQI is represented as the sum of the entropy weights and the quality rating scale and it is articulated as E-WQI = $\sum (W)_j * (U)_j$ where $(U)_j$ for each parameter is given as the ratio of the monitored value $(I)_j$ to its standard value $(S)_j$: $U_j = (I_j/S_j) * 100$ and weight of each parameter is estimated as $(W)_j = (1 - E_j)/\sum (1 - E_j)$, whereas $(E)_j$ refers to overall sample entropy information. The water quality classification scale as recommended by Wu et al. (2011) indicates waters with E-WQI < 50 to be of excellent class, score between 50 and 100 as good, a value between 100 and 150 as "medium", 150 to 200 as "poor" and finally, E-WQI > 200 as "extremely poor".

In addition to entropy WQIs, researchers have assessed the possibility of multi-objective decision-making techniques in river restoration initiatives with regard to demand response, redressing management, changing WQI ranking, and renewable energy sources. The cornerstone of COPRAS and GRA is an indicative of information entropy, and the entropy technique leads to the derivation of criterion weights (Bharat and Bose 2023). It is noted that for rivers to which a ranking based on pollution level, was to be applied, the parameters (criteria) and sample sites (alternatives) were stipulated (Fan et al. 2022). The weight of the water quality parameter was assessed using methods for Shannon entropy (Kumar et al. 2022). This makes it necessary for the decision-making processes to prioritize decisions during emergency situations in addition to providing an overall ranking of the locations by accounting for heavy metals and physicochemical factors (Liu and Ye 2023). Therefore, it acts as a useful tool for decision-making and can be used to apply a set of equations proposed by Saravanan et al. (2023). The interpolation of different experimental data to produce thematic and spatial maps is made easier by these two processes. Therefore, it also permits the statistical construction of a connection for the purpose of providing a simple visual summary of the area's surface water quality.

1.5 Results and Discussions

Surface water resource management must be sustainable in order to support human health, ecological diversity, and economic expansion. This study makes use of cutting-edge methods and scientific ideas. Arc GIS 10.5 software was used to create the geodatabase, and the IDW method was applied to the spatial interpolation. The parameter namely, pH evaluates the water's acidity and alkalinity. The water pH impacts biotic activity, and the process of photosynthesis also produces a high pH. This implies that there are a lot of active activities in the studied area. Additionally, it is observed to decrease with an increase in thermal input. In the study area, pH levels (7.7–7.9 mg/l) of most of the samples are within the range of 6.5–8.5 and it appears to be alkaline, and the reported pH value was higher than the potability range that the WHO guidelines advised. Afterwards, water becomes more acidic as a result of the reduction in the hydrogen–oxygen reaction brought on by the release of hydrogen ions as temperatures rise.

On the other hand, a decrease in temperature increases the DO level in water and strengthens the hydrogen–oxygen connection. Optimum amount of DO (7.2–7.8) is being reported and found healthy in all sites. BOD indicates the amount of garbage in the water that is biologically degradable. Aquatic life may be wiped out by high BOD levels in water because of competition for DO. Moreover, DO is used by microorganisms during the biodegradation process, which could change the water's oxygen content. When compared to the WHO norm of 5 mg/L, where they start to pose a threat to aquatic life due to insufficient O_2 availability, the obtained BOD values (1–2.4) are incredibly low. This could indicate less organic waste, and the aquatic life is dependent on the environment of rivers.

Coliform level (1,212–42,529 MPN/100 ml) was within the threshold limit of WHO (5000 MPN/100 ml) except ST-8, 9 and 19. This is a result of the pollution of river water by household trash, industrial waste, and agricultural runoff, which are signs of dangerous toxins in mineral form. In addition, the effects of human activity are layered in certain places, especially those with high levels of human activity.

TDS and TSS might indicate the degree of water contamination. Water conductivity and turbidity may be impacted by a steady rise in their concentrations. When exposed at higher levels, it may have negative health effects on the population that is most at risk. The TDS in the present investigation ranged from 82–13,230 mg/l, whereby the majority of sampling locations fall within the WHO-recommended limit (100 mg/L) except ST-9. Since ST-9 lies 85 km upstream (U/s) from the sea, the high TDS concentration there is the result of sea water intrusion during high tide. TSS primarily comes from open mining regions and is created during excavation, surface runoff, dust, and soil. The recorded values (28.6–74.9) are within the permissible limit (100 mg/l). Water's ability to neutralize acid is referred to as alkalinity or buffering capacity. The value (70–100 mg/L) is within the allowable threshold of WHO (200 mg/L). Because of this, TDS and TSS measurements in the research region met the desired standards. The relatively low results also suggest that the river is less polluted and hence receives comparatively fewer inorganic and organic inputs.

COD quantifies the rate at which organic and inorganic compounds in water dissolve or get suspended by chemical reaction. The values (6–21) were observed and fall under WHO limit of 30 mg/L. The results may suggest that the river's higher temperature and increased rates of biological activity are aggravating organic pollutants.

Soil bacteria mineralize atmospheric nitrogen, which is then transformed into ammonium in an aerobic environment. Nitrifying bacteria then aid in the production of nitrate from ammoniacal nitrogen. NH_3-N varied according to concentrations of 0.5–1.9 mg/L respectively, and exhibits under the desirable threshold of 2 mg/L. Free NH_3 originated from agricultural areas as a result of plant nutrients' and nitrate fertilizers' leaching processes. The concentration varied as (0.02–0.06). Maximum allowed limit is taken as 2 mg/l as per WHO standards.

TKN content in the river stretch was found to vary from (3–11 mg/l) which falls above the permissible limit of WHO in most sites. The maximum value was reported at ST-9 which is near the solid waste dumping and industrial sectors. Surface water contamination resulted from uncontrolled seepage of waste water from industrial and agricultural processes as well as infiltration of urban storm water.

EC quantitatively evaluates the water's capacity to conduct electricity. A high EC measurement has no known effects on public safety, although it may affect the cost of treating and mineralizing water. The EC values all fall in the range 138–7779 μS/cm, their high diversity at Paradeep (7779 > 2250 μS/cm (WHO)) indicating, it is thought that the local concentrations of dissolved minerals depend on the land use and human activity in the watershed. It also raises the possibility that the catchment region may be heavily contaminated with dissolved ions from home and agricultural wastes. Additionally, because of the high concentration of dissolved ions released into the lake, residential and chemical runoffs from farms and different residences may be contributing factors to the potential cause of enhanced water conductivity.

SAR is a criterion that can be used to assess the dangers associated with alkaline soil; a high value reduces soil permeability. Hence, sodium ions are products of weathered silicate rocks. Additionally, it calculates the sodium hazard that irrigation water contaminated with sodium ions may pose. Because they alter the permeability, aeration, texture, and alkalinity of the soil, irrigation water contaminated with sodium ions may slow down plant growth. The value available in the range of 0.4–16.5, making them excellent except ST-9 (16.59), indicating that the water supply is unfit to irrigate farmlands in the catchment area.

Magnesium and calcium are necessary components in water when present in the right amounts. They frequently regulate the hardness of water, and if taken in excess, they can cause stomach problems and scaling in people. Hardness was reported with the maximum and minimum value being 51 mg/l and 2195 mg/l respectively. The hardness showed a slight decrease along the course of river with exceptions near ST-9 (2195 mg/l). Water hardness may have its origin in fertilizer-contaminated agricultural farmlands' runoff. At ST-9, the water's hardness surpasses permitted thresholds, resulting in boiler and pot scaling as well as renal disease in people. Anthropogenic activities such as seepage beds, septic tanks, home or municipal sewage, and nitrogenous waste are the sources of boron pollution. Possible causes include human activity, animal excrement, and nitrogen-based fertilizers oxidizing due to phytoplankton activity and microbial nitrification. It ranged from 0.03–0.5 mg/l, which was below the permissible limit by WHO (2 mg/l).

Fe^{2+} element in water can cause the water to turn colored, which starts the process of systemic sedimentation and corrosion. Increased concentration can have more detrimental effects on human consumption and induce encrustation in the water supply infrastructure. This parameter can have a detectable taste in drinking water, and at very high concentrations, it can have a laxative impact on many users. It ranged between 0.6–2.5 mg/l. Since the value is below the allowable level of 3.0 mg/l, there is no health risk. The cation analysis demonstrates that the order of significance is varied in a sequence of $Fe^{2+} > B^{+}$.

The permeability and porosity of the soil are also crucial in the development of Cl^- concentrations. However, excessive Cl-levels render water unpleasant to drink and inappropriate for watering livestock. Apart from ST-9, the Cl^- concentrations range from 9 to 4904 mg/l, suggesting that they are well under the drinking guidelines (250 mg/l). This is due to the presence of irrigation drains, waste outflows, and domestic sewage. Drinking water with a high SO_4^{2-} concentration shows that the quality of the water is deteriorating and Gypsum is a significant factor in the elevated sulphate levels found in many aquifers worldwide. It is frequently produced when sulfur-containing minerals weather oxidatively. However, the concentrations (4–376) were below the permissible limit (200 mg/l as per WHO) in all the investigated samples except for ST-9 (376 mg/l). Leachate and domestic wastewater contain both organic and inorganic pollutants, which are the primary sources of contamination.

Generally, main sources of F^- include municipal solid waste dumps, industrial garbage, and seepage of untreated sewage water into rivers. Uncontrolled agricultural practices involving the use of agrochemicals and fertilizers throughout the watershed may have an impact on the primary components. The F^- content (0.26–1 mg/l) was below the guideline value designated by WHO (1 mg/l).

NO_3^- is present in waste water, domestic waste, agriculture, industry, and landfills. This may be the result of anthropogenic factors like mixing chemicals and animal wastes during irrigation or leftover waste water in the area, as well as soil leaching. However, values (1–2.7 mg/l) showed that all parameters were with the permitted limits (45 mg/l). As a result (Fig. 1.2), the typical anion concentration in water indicated that chloride largely dominates followed by sulphate, nitrate and finally, fluoride, in the order of $Cl^- > SO_4^{2-} > NO_3^- > F$. We may therefore conclude that, apart from ST-(8) and (9), the water quality in this area is adequate for home, drinking, agricultural, and industrial applications after comparing the amounts of specific water quality parameters present at the site area with WHO approved values.

In spatial comparisons of locations (Fig. 1.3), the EWQI approach has been used to rate the overall water quality. This method, which involves classifying and modelling environmental data, is claimed to be the most effective in preventing incorrect interpretation of quality monitoring data.

In the study area (Table 1.1), it ranged from 14.6–1065.2, showing excellent to extremely poor class. According to EWQI classifications, 84.20% of the samples are categorized as excellent, 10.53% of the samples are poor and 5.26% of the points belong to extremely poor, hence unsuitable for drinking. The score of E-WQI in ST-9 (1065), ST-8 (196) and ST-19 (152) indicated that the water quality in these provinces was classified as poor/extremely poor. Also, higher value of entropy has been determined to be primarily responsible for SO_4^{2-}, Cl^-, EC, SAR, TDS, TH, SAR, TKN and TC. The findings show that urbanization, continuous irrigation with fertilizers, enrichment of Cl^- and SO_4^{2-}, changing crop patterns, monotonous crops, rainfall influence, indiscriminate use of wastewater in agriculture, weathering of rock, and dominance of evaporation are all factors contributing to poor water quality. It's crucial to remember that MCDM approaches for independently analyzing the weight of the criteria can streamline the advantages and disadvantages of the procedures used to create the EWQI for assessing water quality.

The overall ranking and performance score (PS) of all the sampling points in the present study, given by the COPRAS and GRA methodology, has been shown in the Table 1.2.

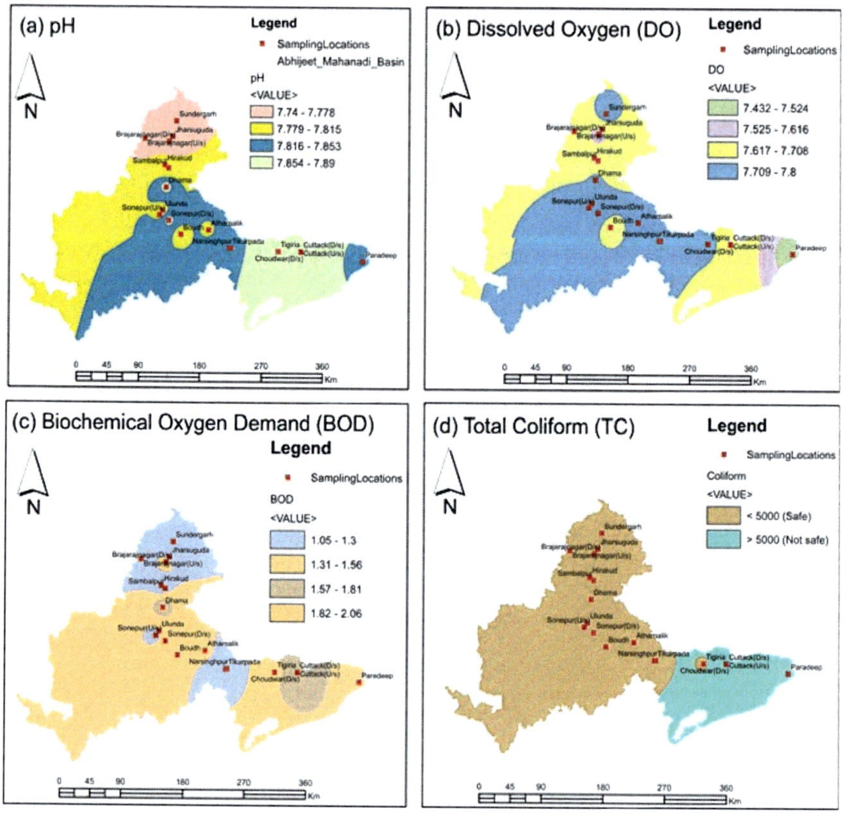

Fig. 1.2 Spatial trend map of all 20 indicators

By assigning them an overall score of 1 or 2, respectively, these two methodologies likewise placed these areas among the sites with the highest levels of pollution. COPRAS and GRA approach, however, clearly determines the relative pollution level by providing their overall ranks in such a complex scenario. It is noteworthy that the sites (9), (8) and (19) were identified as relatively most polluted site in compared to other locations which depicts high disorder indices around these areas during the monitoring period of 2015–2023, that caused greater variability (Figs. 1.4 and 1.5). High fluctuation in water quality was a result of dumping, being close to a road, and sewage effluents at these places.

The present EWQI, COPRAS and GRA investigation successfully locates areas where the cumulative effects of anthropogenic and natural activities have had the greatest impact on the quality of the water. Thus, areas with low water quality need to be monitored frequently. Such prime places for the beginning of restoration operations may also allow for the assessment of substantial effects of dilution, changes in water depth, and runoff (Das, A 2023).

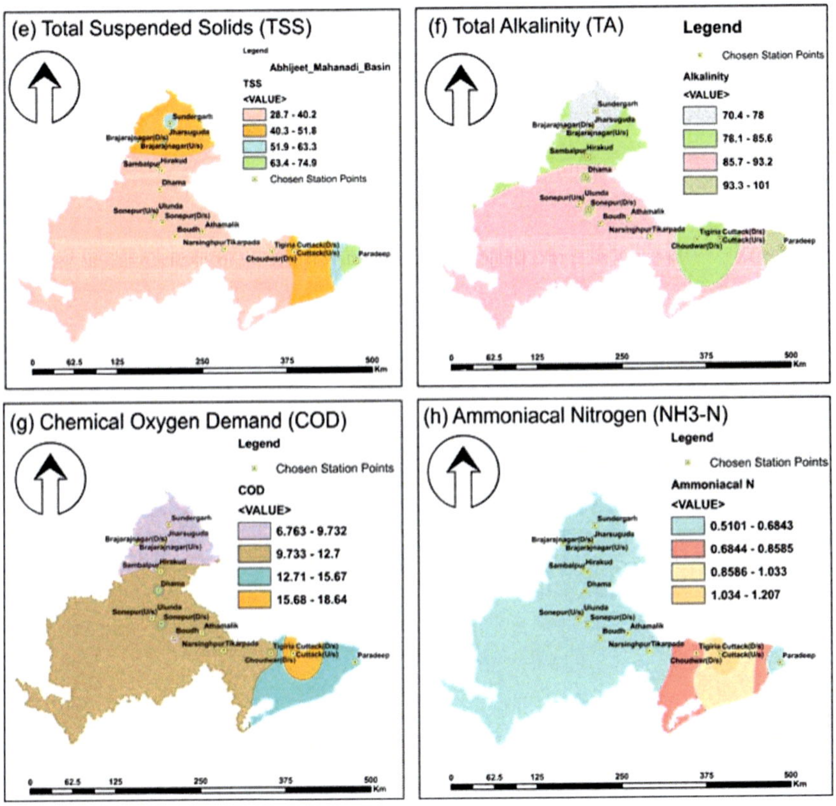

Fig. 1.2 (continued)

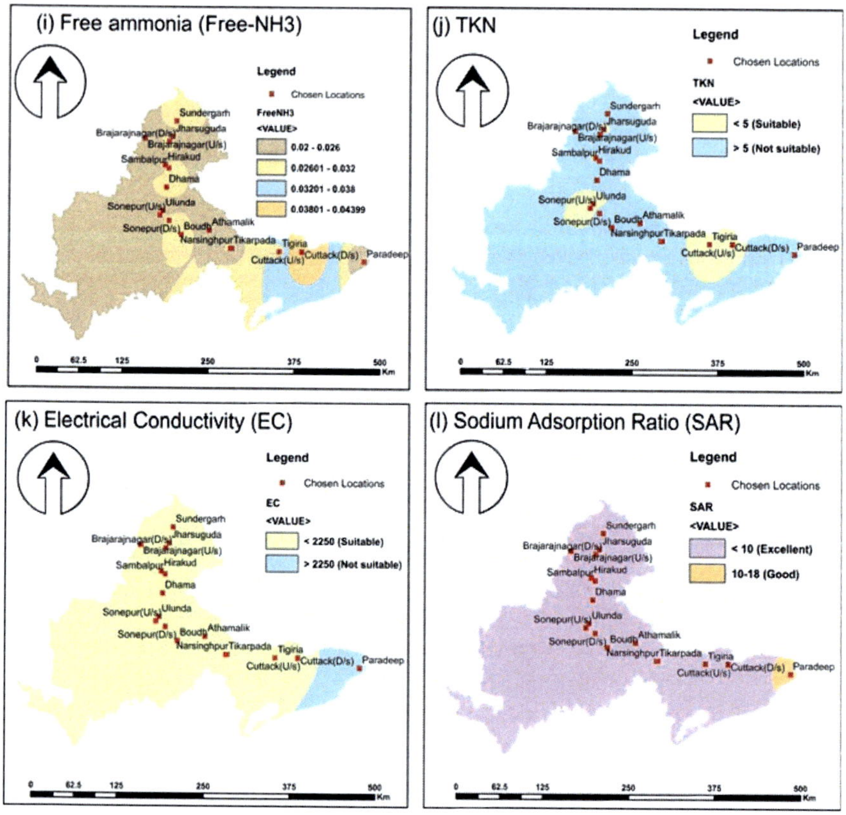

Fig. 1.2 (continued)

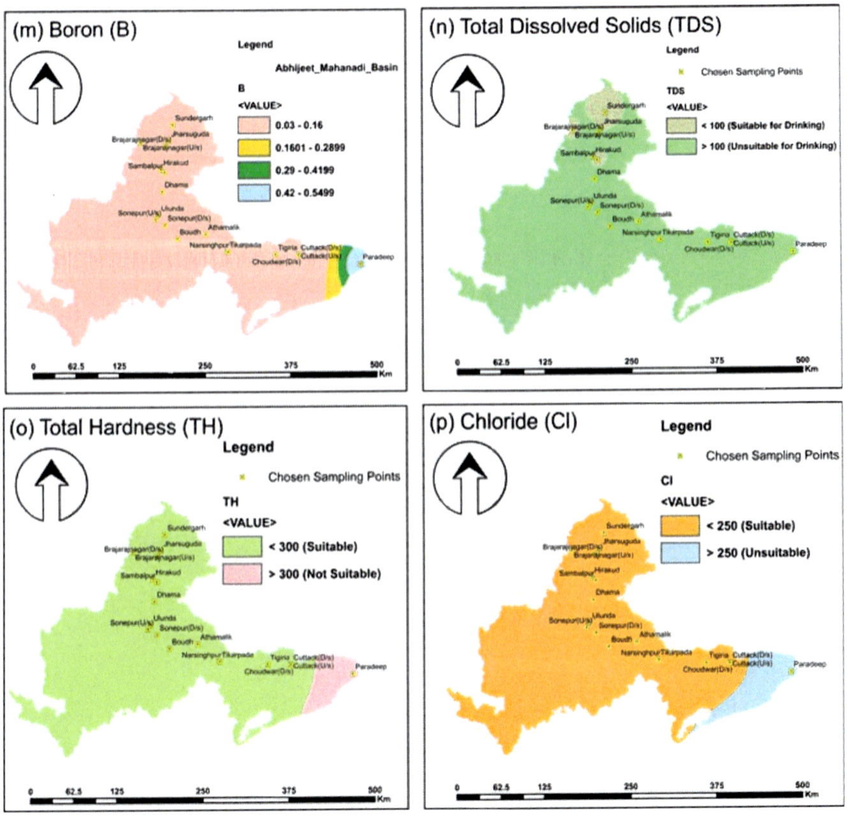

Fig. 1.2 (continued)

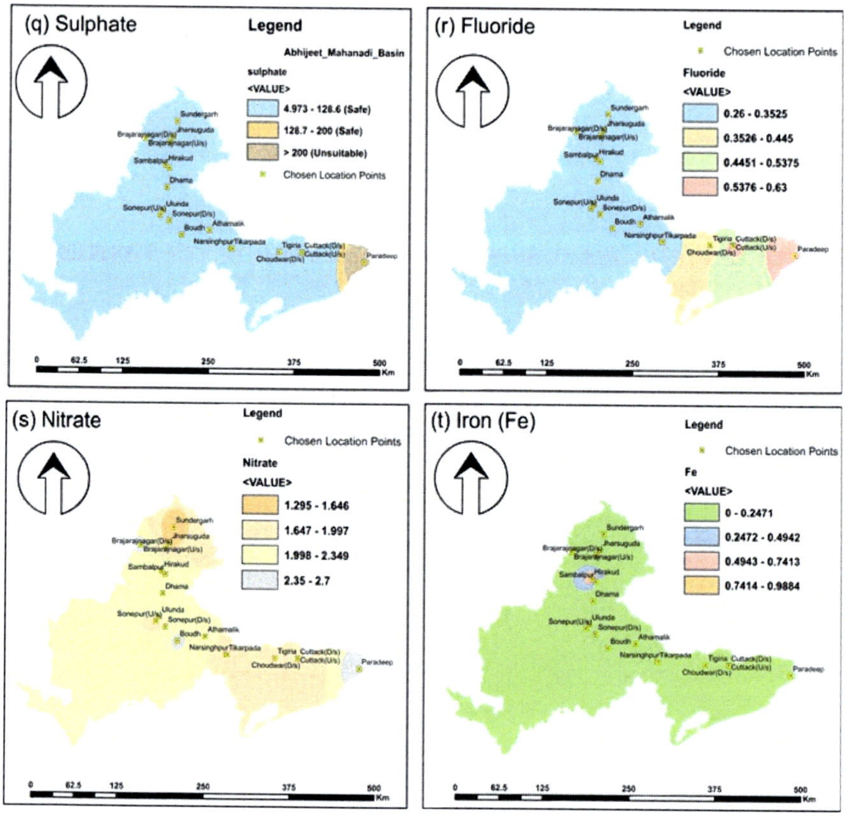

Fig. 1.2 (continued)

It is expected that the research findings will be a great help to policy makers associated with water quality safeguard organizations. In the present study, Entropy and decision-making tools make it simple to determine the sampling locations' relative pollution levels in relation to the drinking water quality requirement. However, it becomes extremely difficult to locate appropriate places for the intake structures of irrigation canals for diverting the river water because there is no precise system for classifying irrigation water quality with respect to allowable limitations. In these situations, incorporating important irrigation and heavy metal characteristics into MCDM methods for inquiry may provide a workable answer. In addition, future research can also consistently analyze the physicochemical parameters' sensitivity.

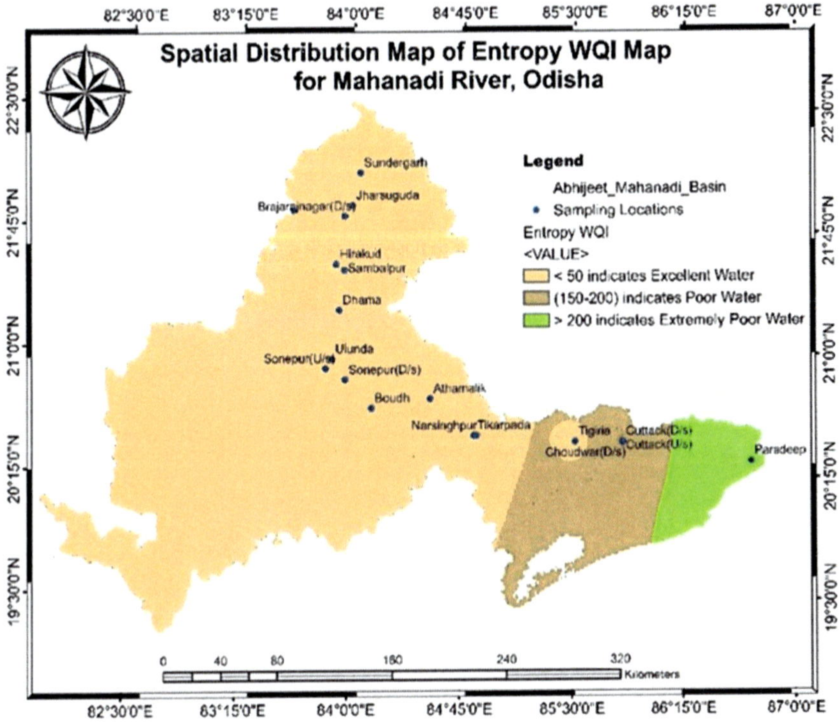

Fig. 1.3 Distribution of geo-spatial map of EWQI

1.6 Conclusions

This experiment is intended to assess the quality of the surface water within the urban reach of Mahanadi River Basin, Odisha, which is located in the State of Odisha and is experiencing a high pace of population growth and uncontrolled urban and industrial activity expansion, that could cause the surface water quality to progressively deteriorate. The likelihood of drinking was assessed by E-WQI and MCDM method, following WHO standards, and basically, focuses on 20 samples of surface water taken over the course of 8-year (2015–2023). Results show that TKN and TC are found to be higher than the threshold limit in all sites. The spatial variation was demonstrated by the ArcGIS program-based on spatial analyst approach. The E-WQI coupled with MCDMs jointly exhibit poor quality of 15.79% in Paradeep (ST-9), Cuttack D/s (8) and Choudwar D/s (19), in total 3 locations, demonstrating efficient ionic leaching, overuse, and anthropogenic activities from the discharge of effluents from industrial, agricultural, and home applications. In order to conserve the delicate environment, good planning is crucial in this project as the water quality

Table 1.1 E-WQI values and its viability for drinking in the investigated study area

Site No.	E-WQI value	Water type
ST-1	15.7	Excellent
ST-2	18.4	Excellent
ST-3	16.2	Excellent
ST-4	19.9	Excellent
ST-5	18.6	Excellent
ST-6	19.5	Excellent
ST-7	17.6	Excellent
ST-8	196.0	Poor
ST-9	1065.2	Extremely Poor
ST-10	15.0	Excellent
ST-11	15.1	Excellent
ST-12	14.6	Excellent
ST-13	16.9	Excellent
ST-14	20.0	Excellent
ST-15	16.3	Excellent
ST-16	18.4	Excellent
ST-17	19.5	Excellent
ST-18	17.9	Excellent
ST-19	152.0	Poor

in the aforementioned locations is slowly deteriorating to an alarming level. The suitable zoning map of E-WQI, GRA and COPRAS generated through GIS ascertains the applicability of coupled models and necessity of regional scale analysis.

Table 1.2 Priority ranking of regions

	COPRAS		GRA	
Site No.	PS (Ui)	Ranking	PS (Ri)	Ranking
ST-1	2.18	16	0.0169	18
ST-2	2.52	9	0.0171	9
ST-3	2.20	15	0.0170	17
ST-4	2.66	5	0.0173	5
ST-5	2.48	11	0.0171	12
ST-6	2.49	10	0.0171	11
ST-7	2.43	12	0.0172	7
ST-8	6.23	2	0.0194	2
ST-9	100.00	1	0.0469	1
ST-10	2.08	19	0.0171	13
ST-11	2.09	18	0.0169	19
ST-12	2.09	17	0.0170	15
ST-13	2.28	14	0.0170	14
ST-14	2.62	7	0.0173	4
ST-15	2.34	13	0.0170	16
ST-16	2.69	4	0.0172	8
ST-17	2.60	8	0.0171	10
ST-18	2.65	6	0.0172	6
ST-19	3.87	3	0.0185	3

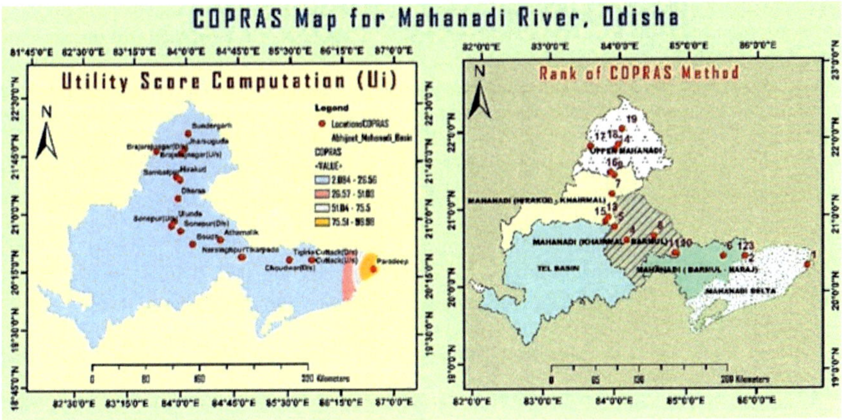

Fig. 1.4 COPRAS investigation map

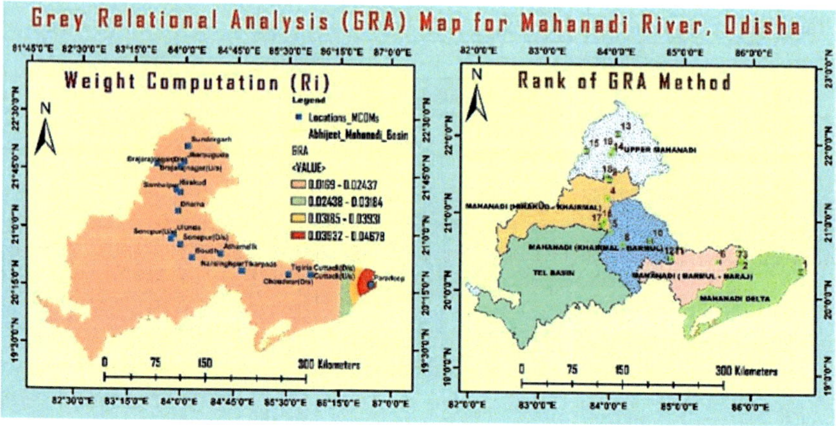

Fig. 1.5 GRA investigation map

Acknowledgements The authors would like to express their sincere gratitude towards State Pollution Control Board, Odisha for providing the online data of water quality for river Mahanadi. I also would like to convey my gratefulness towards C.V. Raman Global University in Odisha, India, for providing the scientific team with the necessary research labs.

References

Agarwal N, Tayal DK (2023) A new model based on the extended COPRAS method for improving performance during the accreditation process of Indian Higher Educational Institutions. Comp Appl Eng Educ 31(3):728–754

Allafta H, Opp C, Patra S (2021) Identification of groundwater potential zones using remote sensing and GIS techniques: a case study of the Shatt al-Arab Basin. Remote Sens 13:1–28. https://doi.org/10.3390/rs13010112

APHA (2012) Standard methods for the examination of water and wastewater, 22nd edn. American Public Health Association, Washington, DC. Accessed 24 Feb 2021

Bharat N, Bose PSC (2023) Wear performance analysis and optimization of process parameters of novel AA7178/nTiO2 using ANN-GRA method. In: Proceedings of the Institution of Mechanical Engineers, Part E: Journal of Process Mechanical Engineering, 09544089231156074

Chang KH, Chung HY, Wang CN, Lai YD, Wu CH (2023) A new hybrid Fermatean fuzzy set and entropy method for risk assessment. Axioms 12(1):58

Das A (2022a) Multivariate statistical approach for the assessment of water quality of Mahanadi basin, Odisha. Mater Today Proc 65:A1–A11

Das A (2022b, December) Surface Water Quality Modelling Using Water Quality Index (WQI) and Geographic Information System (GIS) on the Mahanadi Basin, Odisha. In: International conference on water technologies. Springer Nature Singapore, Singapore, pp 21–46

Das A (2023a) Modelling of surface water quality and spatial mapping in Mahanadi Basin, Odisha. Journal of Mines, Metals & Fuels 71(9):1155–1173

Das A (2023b) Identification of surface water contamination zones and its sources on Mahanadi River, Odisha Using Entropy-Based WQI and MCDM Techniques. Engineering Research Transcripts 4:67–92

Das A (2023c) Anthropogenic effects on surface water quality assessment in Baitarani River Basin, Odisha Using GIS and MCDM Techniques. Eng Res Transcripts 5:37–64

Das A (2023d) Assessment of potability of surface water and its health implication in Mahanadi Basin, Odisha. Mater Today Proc

Das M, Parveen T, Ghosh D, Alam J (2021) Assessing groundwater status and human perception in drought-prone areas: a case of Bankura-I and Bankura-II blocks, West Bengal (India). Environ Earth Sci 80(18). https://doi.org/10.1007/s12665-021-09909-8

Dibs H, Ali AH, Al-Ansari N, Abed SA (2023) Fusion Landsat-8 thermal TIRS and OLI datasets for superior monitoring and change detection using remote sensing. Emerging Sci J 7(2):428–444. https://doi.org/10.28991/ESJ-2023-07-02-09

Egbueri JC (2022) Predicting and analysing the quality of water resources for industrial purposes using integrated data-intelligent algorithms. Groundwater Sustain Dev 18:100794

Fan J, Han D, Wu M (2022) T-spherical fuzzy COPRAS method for multi-criteria decision-making problem. J Intel Fuzzy Syst 43(3):2789–2801

Goswami T, Ghosal S (2022) Understanding the suitability of two MCDM techniques in mapping the groundwater potential zones of semi-arid Bankura District in Eastern India. Groundwater Sustainable Dev 17:100727. https://doi.org/10.1016/j.gsd.2022.100727

Gupta N, Mathew A, Khandelwal S (2020) Spatio-temporal impact assessment of land use/land cover (LU-LC) change on land surface temperatures over Jaipur city in India. Int J Urban Sustain Dev 12(3):283–299

Ji Y, Sheng Q, Zhu Z (2023) Assessment of ecological benefits of urban green spaces in Nanjing city, China, based on the entropy method and the coupling harmonious degree model. Sustainability 15(13):10516

Kumar M, Singh P, Singh P (2022) Integrating GIS and remote sensing for delineation of groundwater potential zones in Bundelkhand Region, India, Egypt. J Remote Sens Space Sci 25:387–404. https://doi.org/10.1016/J.EJRS.2022.03.003

Liu Y, Ye M (2023) Application and validity analysis of IoT in smart city based on entropy method. Appl Artif Intell 37(1):2166234

Nguyen TG, Phan KA, Huynh THN (2022) Application of integrated-weight water quality index in groundwater quality evaluation. Civ Eng J 8:11. https://doi.org/10.28991/CEJ-2022-08-11-020

Nyakundi R, Nyadawa M, Mwangi J (2022) Effect of recharge and abstraction on groundwater levels. Civ Eng J 8:05. https://doi.org/10.28991/CEJ-2022-08-05-05

Omar PJ, Dwivedi SB, Dikshit PKS (2020) Sustainable development and management of groundwater in Varanasi, India. In: AlKhaddar R, Singh RK, Dutta S, Kumari M (eds) Advances in water resources engineering and management. Springer Singapore, Singapore, pp 201–209

Qadir J, Bhat MS, Alam A, Rashid I (2020) Mapping groundwater potential zones using remote sensing and GIS approach in Jammu Himalaya, Jammu and Kashmir. GeoJournal 85:487–504. https://doi.org/10.1007/S10708-019-09981-5/TABLES/10

Roy S, Bose A, Mandal G (2022) Modeling and mapping geospatial distribution of groundwater potential zones in Darjeeling Himalayan region of India using analytical hierarchy process and GIS technique. Model Earth Syst Environ 8(2):1563–1584

Saravanan V, Ramchandran M, Murugan A (2023) A Study on Alumina Nana Particles Mechanical Properties using the GRA Method. J Mater Charact 2(2):1–8

Shannon CE (1948) A mathematical theory of communication. Bell Syst Tech J 27:379–423

Singh KR, Dutta R, Kalamdhad AS, Kumar B (2020) Review of existing heavy metal contamination indices and development of an entropy-based improved indexing approach. Environ Dev Sustain 22(8):7847–7864

Thacker H, Shah Y, Borah AJ, Jadeja Y, Thakkar M, Bhimani S, Chauhan G (2022) Assessment of groundwater potential zones across Katrol Hill Fault, Kachchh, Western India: a remote sensing and GIS approach. Open J Geol 12(2):111–135

WHO (2017) Guidelines for drinking water quality, 4th edn. World Health Organization, Geneva, Switzerland

Wu J, Li P, Qian H (2011) Groundwater quality in Jingyuan County, a semi-humid area in Northwest China. J Chem 8(2):787–793

Zade N, Avadhoot B, Kumar DP, Pradip S, Robin D (2021) Variability of mechanical properties of cellular lightweight concrete infill and its effect on seismic safety. Nat Hazard Rev 22:4021039. https://doi.org/10.1061/(ASCE)NH.1527-6996.0000501

Chapter 2
Runoff Modelling for the Watershed Using Hydrological Model Swat Under Changing Environment

Nagendra Reddy, Mahesh Kumar Chitrahalli Lingaraju, Shwetha Kotagi Girisha, and Milad Khatib

Abstract Climate change has a significant impact on water resource availability, by which the rainfall patterns also change. In the present study, a hydrological model called Soil and Water Assessment Tool (SWAT) has been used to produce a Rainfall-Runoff modeling for the study area, the Doni River, with a catchment area of 3,399.42 km^2. Discharge data from 1995 to 2014 was used for calibration and validation. The calibration period was considered from 2000 to 2004 and the validation period from 2005 to 2007. The observed and simulated data results showed good model agreement with statistical performance indicators R2 and Nash Sutcliffe efficiency (NSE) values of 0.87 and 0.85 respectively. Thus, Surface runoff was influenced by sensitive parameters such as CN and ESCO, which in turn affect the subsurface parameters. It was observed that the simulated model values were well correlated with the observed values and hence effective use of the SWAT model can be achieved by its proper calibration and validation.

Keywords SWAT · Doni River · Calibration · Validation

N. Reddy
Department of Civil Engineering, Nagarjuna College of Engineering and Technology, Mudugurki, Venkatagiri Kote Post, Devanahalli, Bengaluru 562164, India

M. K. Chitrahalli Lingaraju · S. Kotagi Girisha
Department of Civil Engineering, NITTE Meenakshi Institute of Technology, Yelahanka, Bengaluru 560064, India

M. Khatib (✉)
Engineering Department, LIU, Mouseitbeh, Beirut, Lebanon
e-mail: milad.khatib@liu.edu.lb; milad.khatib@isae.edu.lb

Engineering Department, ISSEA-Cnam, Beirut, Lebanon

Z. Wan (ed.), *Water Resources Management and Water Pollution Control*, Environmental Science and Engineering, https://doi.org/10.1007/978-3-031-53456-0_2

2.1 Introduction

Water is the most important element for the survival of all living things. It is a crucial element for industrial development, irrigation activities, and economic progress. However, due to the expanding population and industrialization, water has become insufficient to meet the growing needs, as well as a scarce resource (Reddy et al. 2021a).

Under the scenario of changing climate, the availability of water has become challenging for irrigation, food security, and the health of the ecosystem (Jatin et al. 2018). Further changing climate does have more effect on water resources and their quality (Chanapathi et al. 2018). Intergovernmental Panel on Climate Change (IPCC) recently released a report that shows an approximately 4 °C increase in global mean temperature by 2100 (Navneet et al. 2017). Further, the studies show anthropogenic activity is the dominant cause of observed warming, since the mid-twentieth century (Yadav and Junaid 2015; Kumar and Jayakumar 2020). In view of this, global warming and its effects on living beings, regional availability of water, agricultural productivity, etc.... have drawn considerable attention (Kannan and Ghosh 2011; Shashikanth et al. 2014; Liu et al. 2016). Transitional climatic patterns in the past, present, and future, as well as their variability on climate variables such as rainfall and temperature, are crucial for analysing and simulating the regional climate change impacts over river basins (Reddy et al. 2021b).

The discoveries of the IPCC, 2001 have proposed that the change in temperature affects the precipitation pattern, which has a direct effect on hydrology, land and water assets, farming practices lastly biological systems and is additionally in charge of expanded outrageous occasions like floods and dry spells (Moriasi et al. 2015). The ongoing changes in the precipitation and outrageous occasions, considered as a reason for worldwide global warming incited climate change, which leads to devastating effects on water resource availability in nature.

Thus, it is unfavourable for agricultural production. Accordingly, a careful study was undertaken for variable rainfall patterns and their influence on shallow water resources. The future influence of climate change has been identified by "India's National Communications" (NATCOM) in 2004. The impacts are as follows; diminished snow spread, influencing snow and frozen river systems, unpredictable storms with genuine consequences for agriculture that rely on rainwater, increased frequency and intensity of floods, and rising water levels inducing population migration along the thickly populated coastlines of the world. The change in monsoon trend is due to varying rainfall patterns, which is due to the effect of global warming in India. This situation leads to potential consequences for a country where 70% of the population is dependent on agriculture. Thus, there is a changing soil erosion pattern due to the effect of climate change.

These consequences have guided the study of the effect of climate change through hydrological models, which are considered as the improvement of a certifiable framework that helps in comprehension, foreseeing, and managing water assets. These models are based on the dominant hydrological process and the "Hydrologic cycle".

Presently there is innovation in technology in the field of Remote Sensing and Geographic Information systems, which helps in managing and modelling water resources.

Changing climate has a significant effect on the hydrological cycle and consequently on the streamflow, to assess the impact of climate change on streamflow, hydrological modeling plays a crucial role. In this study, and based upon the literature review, a semi-distributed hydrological model soil and water analysis tool (SWAT) is used to build the rainfall-runoff relationship over the Doni River.

2.2 SWAT Model Description

SWAT model defined as "Soil and Water Assessment tool" is a physically based, continuous time, semi-distributed model and it is developed by the United States Department of Agriculture (USDA), Agricultural Research Service (ARS) modeling in 1990's". This model aims at figuring out the effects of water resource management decisions and the pollution caused in river basins due to non-point sources. By using this model, water resource managers can predict and assess the influence of water management, sediment, bacteria, algae, and chemical yields in agricultural land in large river basins.

SWAT is a regular time-based model, which operates on a routine time basis. Components of the model are erosion/sedimentation, channel routing, plant growth, weather, soil temperature, hydrology, nutrients, agricultural management, pesticides, and pond/reservoir routing. This model can provide long-term yield prediction of sediment, nutrients, pesticides, and so on. It consists of both physical and empirically based components. It uses data like topography, land use, soil, and weather, which are spatially distributed.

SWAT splits the watershed into various sub-basins for modeling. "Each sub-basin delineated within the model is simulated as a homogeneous area in terms of climatic conditions". Each of these sub-basins is divided into Hydrological Response Unit (HRU) to represent different soils and land use types. Soil, topography, and land use are spatially uniform. For development and calibration, SWAT requires a notable amount of data and empirical parameters. To develop and calibrate models of hydrology and water quality in a watershed, a sizable amount of detailed data and empirical parameters for the climate, soil characteristics, surroundings, plants, and agricultural management methods are needed.

The model output includes all water balance components (surface runoff, evaporation, lateral flow, recharge, percolation, sediment yield, nutrients, and pesticides) at the level of each sub-basin and is available at daily, monthly, or annual time steps. In HRU, the daily water budget is computed depending on runoff, percolation, evapotranspiration, daily precipitation, and return flow from the subsurface, groundwater flow.

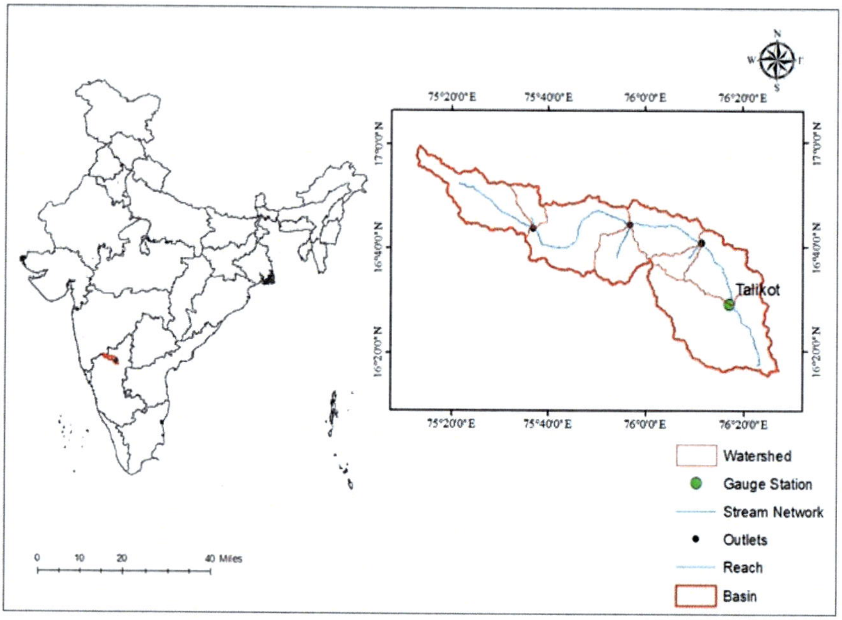

Fig. 2.1 Location of the study area

2.2.1 *Study Area: The Doni River*

The Doni River flows from the eastern side of the Sangli District of Maharashtra State in India (Fig. 2.1). Towards the Eastern side, the river Doni joins Talikot. It is confluence with Krishna River in the Gulbarga District of Karnataka. The river flows with a distance of 176 km with area river basin around 3,399.42 km^2. Doni is a river is geographically located at latitude 16.10° and longitude 74.10°.

2.2.2 *Data Collection*

Different data are required for setting up the model. Discharge data for the Doni River are available from 1979 to 2014. All the data have been downloaded from specified websites as shown in Table 2.1. The land use land cover (LULC), soil, and slope maps were prepared for the study area and displayed (Fig. 2.2).

Table 2.1 Input data

Data collected	Sources
Discharge data	India-wris.nrcs.gov.in
Digital elevation model	Bhuvan.nrsc.gov.in
Land use/land cover, Soil	swat.tamu.edu
Climate variable (e.g., Rainfall, Temperature)	swat.tamu.edu

2.2.3 Model Setup

Using the collected data, a model set-up was been done. A new SWAT project was saved and the following steps were used to come up with the final model run (Fig. 2.3).

2.2.4 Watershed Delineation

It is the first step in setting up of SWAT model. It is a process in which Data Elevation Model (DEM) are loaded, stream networks are produced by using flow accumulation and flow direction. The location of the stream network has been defined by superimposing the stream network dataset into DEM. Hydrological segmentation (or Hydrological Response Units), and sub-watershed boundaries are predicted. The origin of streams was determined by a 3,399.42 km^2 threshold area.

The number of sub-basins, size, and details of the stream network depending on the study area were determined. Sub basin outlet points were defined by default locating the downstream edge for each branch in the stream. The Arc Swat model allows the user to use the watershed delineator tool to add, delete, or modify the outlet points.

After defining the outlet points of streams, the main watershed delineation outlet point was selected for the Talikot gauge station. Furthermore, watershed delineation was defined for the study area, and the watershed delineation was completed. The total number of watersheds in the Doni River basin was 48.

2.2.5 HRU Analysis

The HRU analysis follows the watershed delineation in the parameterization of the SWAT model. This analysis helps in loading the land use, soil, and slope to the watershed. Using the HRU analysis toolbar, land use maps were introduced into the model, and the lookup tables were provided to the model, and the corresponding values were specified. There were five numbers of classes of entered slope. They were as follows; 0–2%, 2–4%, 4–8%, 8–17%, and 17–9999% respectively. After loading each data, reclassifying, and overlaying it, the overlay report will be generated.

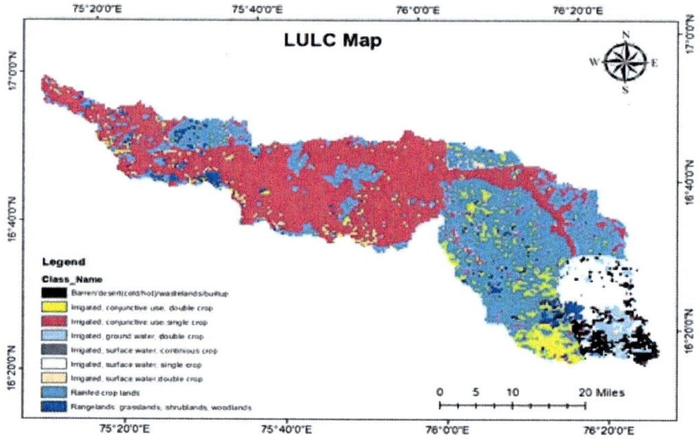

a. LULC Map.

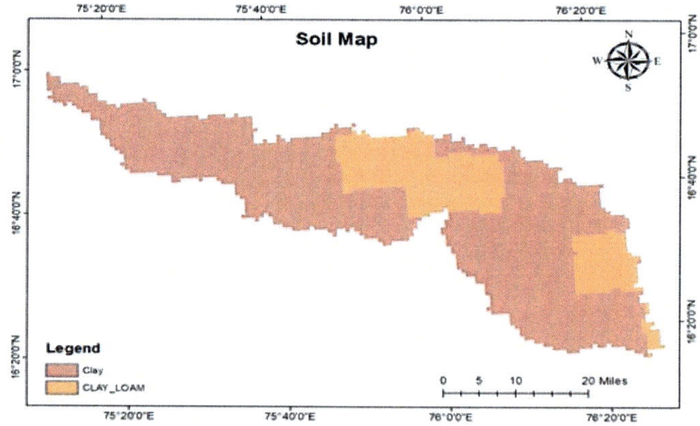

b. Soil Map.

c. Slope Map

Fig. 2.2 Prepared maps. **a** LULC map, **b** Soil map, **c** Slope map

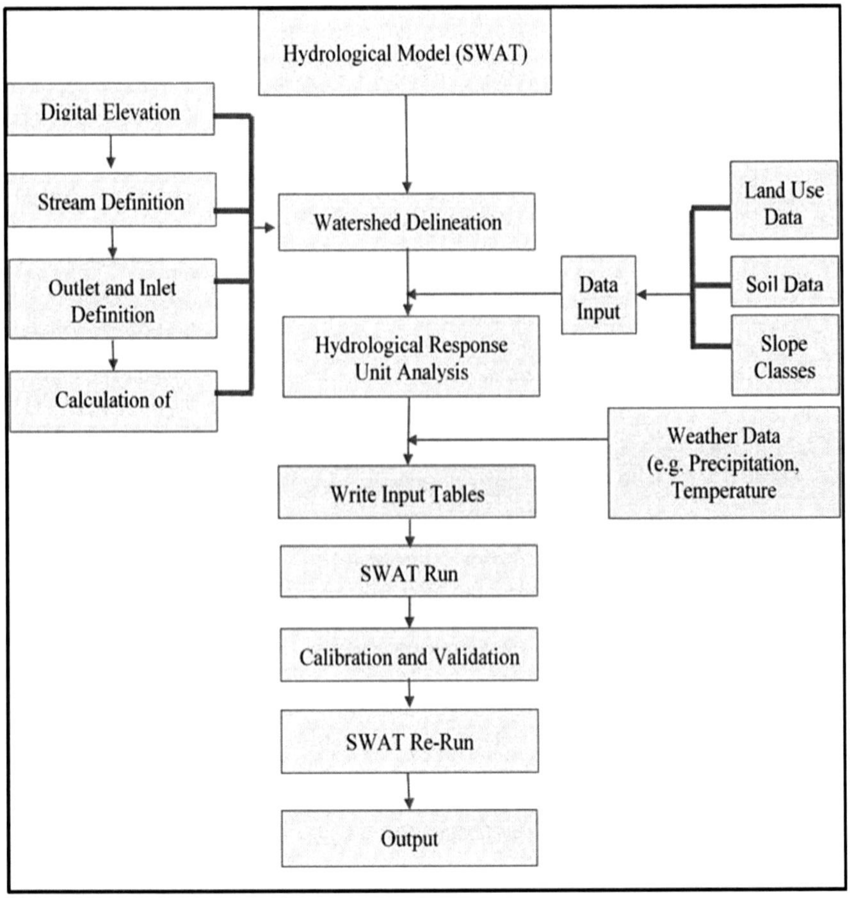

Fig. 2.3 Framework of model

The multiple HRU option was chosen to produce multiple results within each sub-watershed, with a threshold percentage area of land use, soil, and slope as 10% respectively.

2.3 Results and Discussions

Based on sensitive analysis parameters, from the default simulation and by selecting calibrated parameters for the study area Doni River, the obtaining results were validated and discussed below.

Table 2.2 Sensitive parameters ranges in calibration

Sl. No.	Parameter's name	Description	Default range	Calibrated range
1	EPCO	Plant uptake compensation factor	0.01–1	0.01–0.02
2	ESCO	Soil Evaporation compensation	0.01–1	0.01–0.02
3	SURLAG	Surface lag coefficient	0–24	3–4
4	GW_DELAY	Groundwater delay time (days)	0–500	300–400
5	ALPHA_BF	Base flow Recession constant	0–1	0.4–0.5
6	RCHRG_DP	Deep aquifer percolation Fraction	0–1	0.4–0.5
7	GW_REVAP	Groundwater re-evaporation Coefficient	0.02–0.2	0.02–0.03
8	GWQMIN	Threshold depth in shallow aquifer for return low (mm)	0–5000	2000
9	REVAPMN	Threshold depth in shallow aquifer for re-evaporation (mm)	0–1000	400–500
10	CN2	Initial SCS curve number II for moisture	35–98	85–95

2.3.1 Sensitive Parameter Analysis

It refers to the recognition of some parameters that have significant effects on the model. It is the preliminary step to the model calibration. Sensitivity analysis demonstrates the impact that changes to an individual input parameter have on the model response and can be performed using manual calibration and validation. The main objective of the sensitive analysis is to evaluate the output change of the model with respect to the input change of parameters in the watershed. By reviewing the literature study, some sensitive parameters were selected as shown in Table 2.2.

2.3.2 Default Model Run

Hydrologic models were used most frequently to simulate or predict flow either on a continuous basis or for a particular event. In all cases, the model-computed flow was compared with the measured flow. The numerical and graphical performance criteria described below were used in this study for model calibration and validation. The default Arc SWAT model run has been performed for the period of 1979 to

Table 2.3 Calibration and Validation approach over the sub-basin

Area	Approach	Calibration period	Validation period	Discharge gauges location
Doni River	Single-site calibration and validation	2000–2004	2005–2007	Talikot

2014 for the stream flow of the Doni River Basin. The coefficient of determination (R^2) describes the proportion of the total variance in the observed data that can be explained by the model. It ranges from 0.0 to 1.0, with higher values indicating better agreement.

2.3.3 *Calibration and Validation*

In this study, the single-site manual calibration and validation was carried out by considering one site at a time and validating with the same station for different periods within the sub-basin. By carefully ensuring the data availability of monitoring discharge, gauge stations were considered during the calibration and validation phase of gauge station within the study area. The calibration and validation periods considered for the study area and the discharge station used for l calibration over the study area are shown in Table 2.3.

2.3.4 *Performance Evaluation of the SWAT Model*

The performance results of the hydrological model are necessary and are evaluated through the comparison of simulated and observed flow time series of the basin. Generally, the calibrated and validated model performance was evaluated (Table 2.4) with the statistical performance indicators R2 and NSE (Nash Sutcliffe efficiency).

The time series and scatter plots of simulated and observed streamflows at Talikot station are shown respectively (Figs. 2.4 and 2.5) for the calibrated period and the time

Table 2.4 Performance of model using statistical indicators

Statistical Indicator (Moriasi et al. 2015)	Range	Calibration	Validation
$R^2 = \left[\frac{\sum_{i=1}^{n}(Xi-\overline{Xi})(Yi-\overline{Yi})}{\sqrt{\sum_{i=1}^{n}(Xi-\overline{Xi})^2}\sqrt{\sum_{i=1}^{n}(Yi-\overline{Yi})^2}}\right]$	(0, 1)	0.87	0.84
$NSE = 1 - \left[\frac{\sum_{i=1}^{n}(Xi-\overline{Yi})^2}{\sum_{i=1}^{n}(Xi-\overline{Xi})^2}\right]$	$(-\infty, 1)$	0.85	0.79

series. In addition, scatter plot between simulated (Fig. 2.6) and observed streamflows (Fig. 2.7) at Talikot station for the validation period are shown below. Further, the values of R2 and NSE for the calibrated and validated period are presented in Table 2.4.

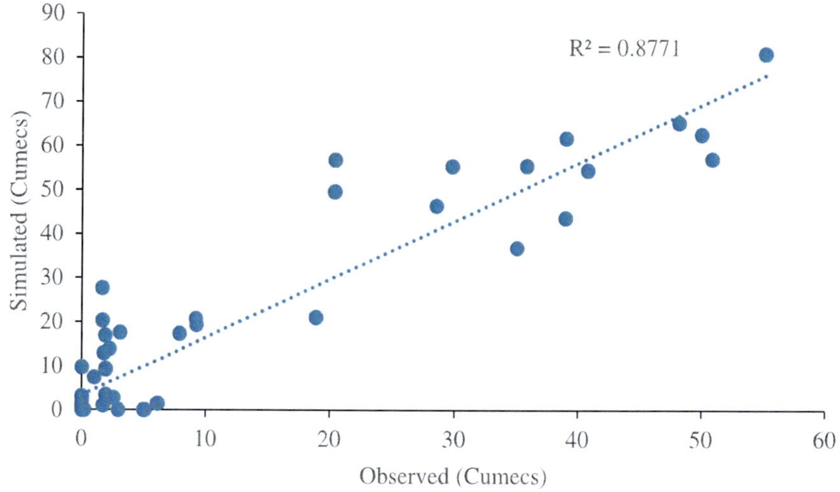

Fig. 2.4 Scatter plot for observed and simulated discharge of calibration period

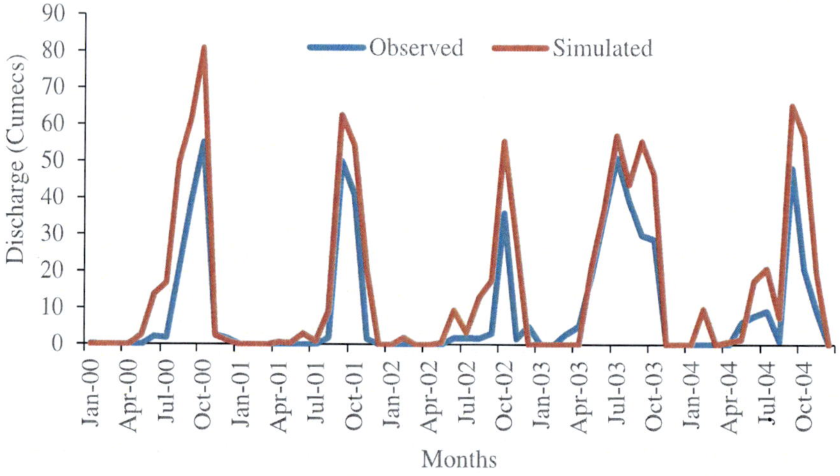

Fig. 2.5 Time series plot of simulated and observed streamflow of calibration period

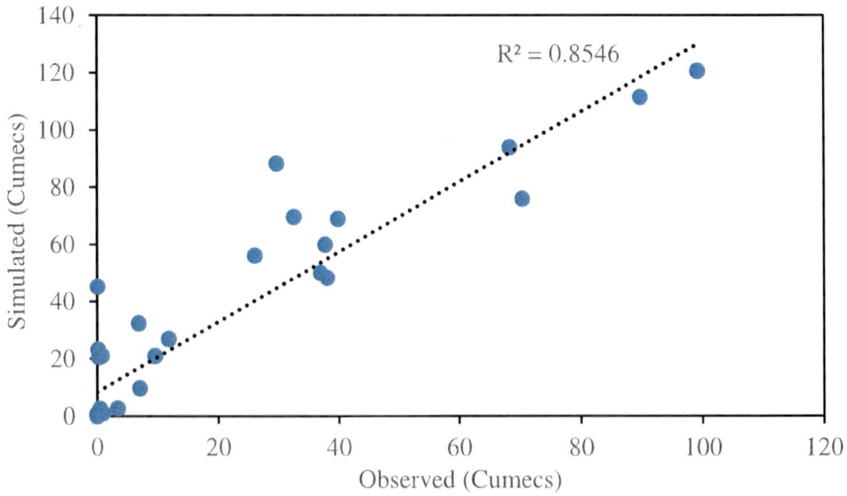

Fig. 2.6 Scatter plot for observed and simulated discharge of validation period

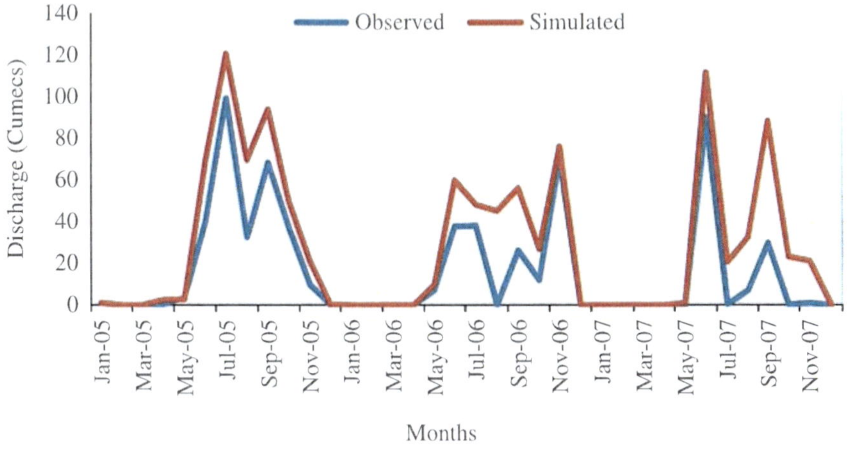

Fig. 2.7 Time series plot of simulated and observed streamflow of validation period

2.4 Conclusions

The model performed well during the simulation of the river according to the available data. This study evaluated and identified sensitive parameters that are affecting the Doni River catchment with an area of 3,399.42 km^2. Simulations are carried out from 1995 to 2014 as the default model runs. Then sensitive parameters have been analysed for the significant model run. Sensitive analysis evaluates the output change of the model with respect to the input change of parameters in the watershed. Then calibration period was considered from 2000 to 2004 and the validation period was from 2005 to 2007.

The results obtained by observed and simulated data showed good model agreement with R2 and NSE values of 0.87 and 0.85 respectively. Runoff modeling helps in understanding how much amount of rainfall turns into rainfall runoff. Surface runoff was influenced by sensitive parameters such as CN and ESCO which in turn impact sub-surface parameters (groundwater) such as GW_REVAP, REVAPMN, GWQMN, ALPHA_BF, GW_DELAY, RCHRG_DP.

The study model shows an increase in precipitation by varying the most sensitive parameters i.e., CN and ESCO. Increased precipitation helps groundwater parameters to recharge the ground. When soil is sufficiently recharged, the remaining rainfall goes as runoff. Hence, sub-surface parameters varied little in their ranges. Using this calibrated model over the selected area, this study can be continued for any future climate change with impact assessment by using different global/regional climate models. It is inferred from the study results that the model-simulated values are well correlated with the observed values and hence effective use of the SWAT model can be achieved by its proper calibration and validation.

References

Chanapathi T, Shashidhar T, Raghavan S (2018) Analysis of rainfall extremes and water yield of Krishna river basin under future climate scenarios. J Hydrol Reg Stud 19(1):287–306. https://doi.org/10.1016/j.ejrh.2018.10.004

Jatin A, Gosain AK, Khosa R, Srinivasan R (2018) Regional scale hydrologic modeling for prediction of water balance, analysis of trends in streamflow and variations in streamflow: the case study of the Ganga River basin. J Hydrol Reg Stud 16:32–53. https://doi.org/10.1016/j.ejrh.2018.02.007

Kannan S, Ghosh S (2011) Prediction of daily rainfall state in a river basin using statistical downscaling from GCM output. Stoch Env Res Risk Assess 25(4):457–474. https://doi.org/10.1007/s00477-010-0415-y

Kumar AU, Jayakumar KV (2020) Hydrological alterations due to anthropogenic activities in Krishna River Basin, India. Ecol Ind 108(105663):1–8. https://doi.org/10.1016/j.ecolind.2019.105663

Liu J, Yuan D, Zhang L, Zou X, Song X (2016) Comparison of three statistical downscaling methods and ensemble downscaling method based on Bayesian model averaging in upper Hanjiang River Basin, China. Adv Meteorol 4:1–12. https://doi.org/10.1155/2016/7463963

Moriasi D, Gitau M, Pai N, Daggupati P (2015) Hydrologic and water quality models: performance measures and evaluation criteria. Trans ASABE 58(6):1763–1785. https://doi.org/10.13031/trans.58.10715

Navneet K, Tischbein B, Kusche J, Laux P, Beg MK, Bogardi J, Bogardi JJ (2017) Impact of climate change on water resources of upper Kharun catchment in Chhattisgarh, India. J Hydrol Reg Stud 13(C):189–207. https://doi.org/10.1016/j.ejrh.2017.07.008

Reddy N, Patil NS, Nataraja M (2021a) Analysis of water balance components of a river sub-basin under future climate scenarios. Sustain Water Resour Manage 7(6):104. https://doi.org/10.1007/s40899-021-00583-z

Reddy N, Patil NS, Nataraja M (2021b) Assessment of climate change impacts on precipitation and temperature in the Ghataprabha sub-basin using CMIP5 models. Mapan 36:803–812. https://doi.org/10.1007/s12647-021-00431-7

Shashikanth K, Madhusoodhanan CG, Ghosh S, Eldho TI, Rajendran K, Murtugudde R (2014) Comparing statistically downscaled simulations of Indian monsoon at different spatial resolutions. J Hydrol 519(3):3163–3177. https://doi.org/10.1016/j.jhydrol.2014.10.042

Yadav BK, Junaid S (2015) Groundwater vulnerability assessment to contamination using soil moisture flow and solute transport modelling. J Irrig Drain Eng 141(7):04014077. https://doi.org/10.1061/(ASCE)IR.1943-4774.0000841

Chapter 3
Preliminary Study of Potential Health Hazard Using Cyprinus Carpio as a Biological Indicator During Construction of Suki Kinari Hydropower Project in Mansehra District, Pakistan

Shan-e-hyder Soomro, Xiaotao Shi, Jiali Guo, Yanqin Bai, Yuanyang Wang, Caihong Hu, Shaista Jalbani, Ao Li, Zhen Yao, and Kang Rui

Abstract Common carp (Cyprinus carpio) is regarded highly beneficial aquatic organism by many Asian as well as some European countries. Due to its remarkable adaptability to diet and habitat, Cyprinus carpio is of exceptional economic importance in freshwater aquaculture in Pakistan. This study aims to evaluate water quality indicators, quantify HMs concentrations in the liver and muscles of Cyprinus carpio. The determination of Fe, Mn, Cu, Pb, and Zn water pollution indices (WPI) and bioaccumulation factors (BAF) were monitored. The results revealed that the water at all sites seemed to have concentration levels of Fe, Pb, and Mn which exceeded the standards set by the World Health Organization in 2011, whereas the concentration levels of Cu and Zn were below the standards. Fe and Cu levels in fish flesh were lower than the FAO, 1992 recommendation, while Mn, Zn, and Pb concentrations were too substantial. To better understand the level toward which metals in the ecosystem cause a risk to aquatic life, the development of frameworks to forecast weather conditions and the impact of habitats on both the quantity and quality of water would benefit from our findings.

S. Soomro · X. Shi · J. Guo (✉) · Y. Bai · Y. Wang · A. Li · Z. Yao · K. Rui
College of Hydraulic and Environmental Engineering, China Three Gorges University, Yichang 443002, China
e-mail: jiali.guo@ctgu.edu.cn

X. Shi
e-mail: fishlab@163.com

S. Soomro · C. Hu
College of Water Conservancy and Transportation, Zhengzhou University, Zhengzhou 450001, China

S. Jalbani
Fisheries and Aquaculture SBBUVAS, Sakrand, Sindh 67210, Pakistan

Z. Wan (ed.), *Water Resources Management and Water Pollution Control*, Environmental Science and Engineering, https://doi.org/10.1007/978-3-031-53456-0_3

Keywords Suki kinari HPP · Cyprinus carpio · Water · Heavy metals · Kunhar River · Pakistan

3.1 Introduction

One of the most important problems in present civilizations is the pollution of the atmosphere from either human and natural systems including hazardous metals (Mitra et al. 2022). These metals can also have toxicant effects on regional fish populations or perhaps harm them if they are substantially accumulated and carried throughout the body across the water, sediment, and aquatic food chain (Raman et al. 2022). In recent years, there has been a rise worldwide in the number of research that investigates the influence of heavy metals (HMs) on aquatic ecosystems (Ustaoğlu et al. 2022). There are approximately 1.5 million fatalities per year due to waterborne disorders; most of such fatalities are among youngsters. On average, 2 million tons of waste are thrown into the world's largest water bodies each day by residential, commercial, and agricultural processes (Gowhar 2018; Shahi 2022). Alkalinity and the total number of soluble solids in the water both increased (Patel and Parikh 2013). Since HMs have such potential to accumulate in bio-systems due to contaminated water, soil, and air origins, HMs pollution of the food supply chain has become a significant issue in recent years (Soomro et al. 2023b). Climate change impacts water temperature, flow regime, channel morphology, and sedimentation, which can be major factors in influencing invertebrates and fish in rivers (Soomro et al. 2022, 2021; Guo et al. 2023). Ecologists all over the globe have already closely observed to this steady increase in metal concentration in water over the past few decades (Li et al. 2022). Unfortunately, there is a lack of data on HMs concentrations as no such research has been performed in Pakistan. The concerns that HMs are inflicting on residents all over Pakistan by drinkable water from the Kunhar River make this research significant.

Bioindicators like fish populations can evaluate the impact of toxic metals on aquatic life (Mamdouh et al. 2022; Soomro et al. 2023c). Moreover, fish can retain metals in their bodies, which they could eventually transmit to humans by consuming these contaminated fish and cause an array of diseases (Huang et al. 2022). Jia et al. (2017) reported that characteristics of the ecosystem contribute to metal accumulation in fish tissues. Many Asian and some European countries see the common carp (Cyprinus carpio) as an extremely valuable aquatic organism. The bioturbation of coastal sediment caused by its grazing on marine habitats impacts aerobic organic matter decomposition and the availability of nutrients in the water column (Hillebrand and Kahlert 2002). Moreover, if there is high level of toxic HMs in the fish body then there are more chances to get effect by consuming that fish as some HMs are also regarded as carcinogenic in nature. Therefore, the purpose of this research is to analyze Cyprinus carpio and water of Kunhar to learn more about the HMs pollution that threatens humans health, local ecology and ecosystem. This study aims to evaluate water quality indicators, quantify HMs concentrations in the liver and muscles of Cyprinus carpio, water, and evaluate the current state of HMs contamination in sediments, the pollution indices; such as water pollution index, metal pollution index, bioaccumulation factor, sediments in water and hazard index.

3.2 Study Area

This study was assessed in the Kunhar River which stretches across Northern Pakistan from 34.2–35.1°N to 73.3–74.1°E; the area of the watershed is 171 km long. It exists in Kaghan Valley and later excursions via Lulusar lake pass through the towns of Batakundi, Naran, Kaghan, Mahandri, Jared, Paras, Kawai, Balakot, Kassi Tarana, and Bisian. It has approximately 2600 km^2 of the drainage area and an elevation from 672 to 5192 m above sea level. Figure 3.1 demonstrates the orographic record of the research region.

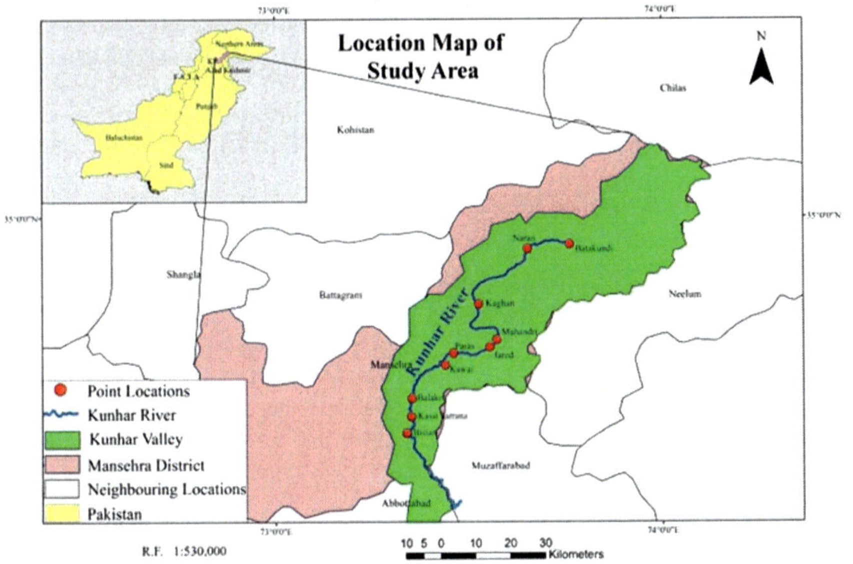

Fig. 3.1 Geographical location and sampling sites in the Kunhar River (Soomro et al. 2023a)

3.3 Materials and Methods

3.3.1 Acquisition and an Analytical Method for Fish Sample

A total of 35 Cyprinus carpio fish were collected from the Kunhar River at the same locations where water samples were taken each season. Overnight, the nets were cast all across the river and left. It was customary to place fishing nets at night (around 18:00 h) and then retrieve them early the following morning (06:00 h.). After collection, they were washed with distilled water to remove any dirt or other possible external items that could taint them before the collection of the muscle samples and were transferred to the laboratory. Before taking each sample, the blade that was used to cut the muscle tissue was cleaned with 1% HNO_3. To calculate the relative amounts of the various HMs that were of interest, one gram of muscle from each fish was used simultaneously from each part. The liver tissues were isolated and washed with distilled water. All samples were kept frozen at −20 °C until further analysis. Before analysis, following the method of (Sattari et al. 2020) the samples were homogenized and processed in a digesting solution consisting of 65% HNO_3 and hydrogen peroxide (H_2O_2), each sample was placed on a plate and heated to a temperature of 200 °C.

3.3.2 Acquisition and an Analytical Method for Water Sample

During the experiment 56 water samples from various points of Kunhar River were collected between March and September (summer, wet season) and November 2021 and (winter, dry season) February 2022; also from main segment of water 35 samples were collected, the remaining 21 samples were collected from significant tributaries and anabranches of the Kunhar River. The water specimen was taken in Polyethylene terephthalate sterile (PET) and were sealed with cover to avoid contamination. In order to prevent bacterial activity from changing the organic content, the specimens were added to the mixture with nitric acid (HNO_3) and then transported in an ice bucket to conducted the analysis. During experiment the analysis was conducted using AR standard quality chemicals that are suitable for analytical testing reagents and all glasswares were properly sterilized before use. Following the method of our recent published article the total dissolved solids (TDS), turbidity, total suspended solids (TSS), total hardness (TH), and biological oxygen demand (BOD) ratio in the river samples were tested (Soomro et al. 2023b) HMs including Fe, Mn, Zn, Cu and Pb were determined in the remainders from acid breakdown and acidic water samples using the ICP-OES method (iCAP 7000, Thermo Scientific, UK) (Amarasinghe 2020).

3.3.3 Water Quality and Health Risk Assessment

Water pollution index (WPI) was determined using the method described by (Hossain and Patra 2020). The method of Burkhard et al. (2021) was followed for calculating the bioaccumulation factor (BAF) in muscles and the liver (Burkhard 2021). It was proposed that the water in the Mansehra district (Khyber Pakhtunkhuwa) area be given a pollution load index (PLI) (Saddique et al. 2018). According to (Isa et al. 2015; Ai et al. 2022) assessment of health risk was done.

3.3.4 Statistical Analysis

Every experiment was repeated three times. The results were statistically analyzed in WINDOWS 2010. Descriptive statistical analysis and ANOVA were utilized to examine the repetitions and evaluate the variations of statistical differences of samples (Mertler et al. 2021).

3.4 Results and Discussion

3.4.1 Assessment of Water Quality

Each metal's contribution to water contamination was determined separately (Fe, Mn, Cu, Pb, and Zn). As can be seen in Fig. 3.2, the sites have been contaminated with a wide variety of HMs. Almost everywhere we looked, the effects were greater than 5, indicating a severe problem with HMs contamination of the aquatic ecosystem. Fe and Pb at site III were in the range of 3–5. When the Water Quality Index (WPI) is less than 1, as reported by (Hossain and Patra 2020) in water samples, this means that no pollution is observed at this site; when the WPI is between 1 and 2, the site is only marginally impacted through contamination. The concentration level of HMs in water sample of Kunhar River watershed mentioned in Fig. 3.2.

3.4.2 Ratio of Metals in Fish Flesh

The concentration of HMs in Cyprinus carpio muscle and liver shows in Fig. 3.3. Fe was highest at site-1 during spring and Mn at site II during spring. Zn was 45.39 μg/g dry wt at site II in spring, Cu was 5.87 at site I in summer, and Pb was 3.41 at site VII in winter. In liver, Fe, Mn, Zn, Cu, and Pb reached 83, 49.37, 57.36, 7.28, and 5.47 μg/g dry wt, respectively. Muscles accumulate HMs less than livers due to poor binding protein and enzymatic activity, hence HMs accumulation is less than in

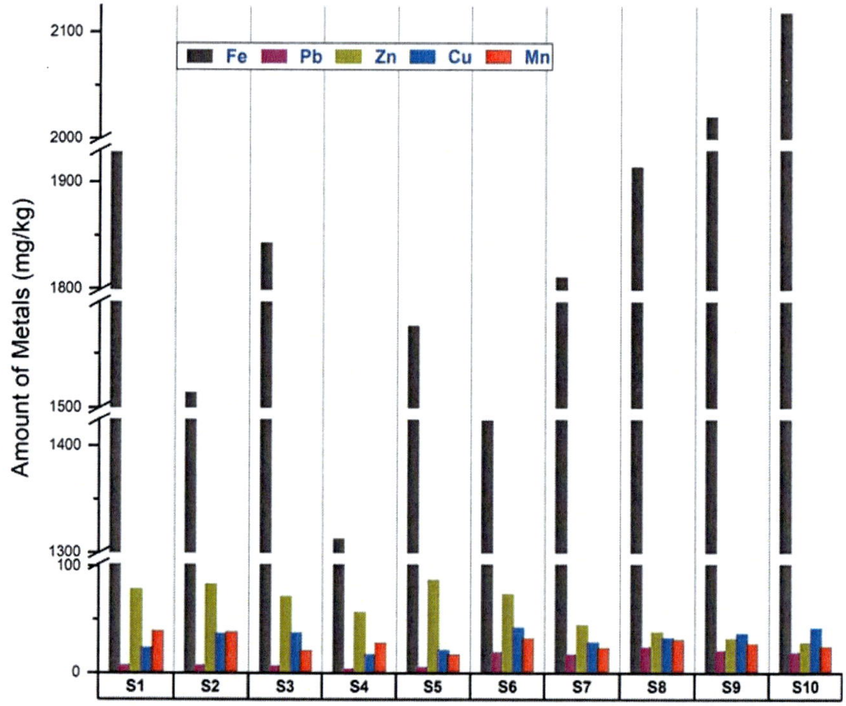

Fig. 3.2 Concentration of HMs in the Kunhar River

the liver (Qadir and Malik 2011). According to, the results indicate that Fe and Cu contaminants are below the acceptable standards, but Mn, Zn, and Pb exceed these limits in fish muscle (Ahmed et al. 2016; Liu et al. 2020).

3.4.3 Bio-Accumulation Factor

HMs bio concentration indices of Kunhar River fish liver and flesh were estimated (Fig. 3.4 and Fig. 3.5). Optimum levels for Fe were observed to be (122.60), and for Mn, values were found to be (1660.00) With in summer, site II recorded high amounts of both Zn and Cu. Apart from Pb, that was also identified (211.11) at site I in the summer. Liver statistics were found to be significantly greater than muscle values, although both sets of findings maintained the same common pattern. Site II had a high portion of Fe, Mn, Zn, and Cu, while site I had the highest concentration of Pb. The ability of different tissues to retain metal fluctuated, with liver having a greater concentration of metals than flesh. The liver's metabolism function may help with metal bioaccumulation (Rajeshkumar and Li 2018; Vinodhini and Narayanan 2008), indicating the chemical reactions that occur inside an alive organism. Natural protein

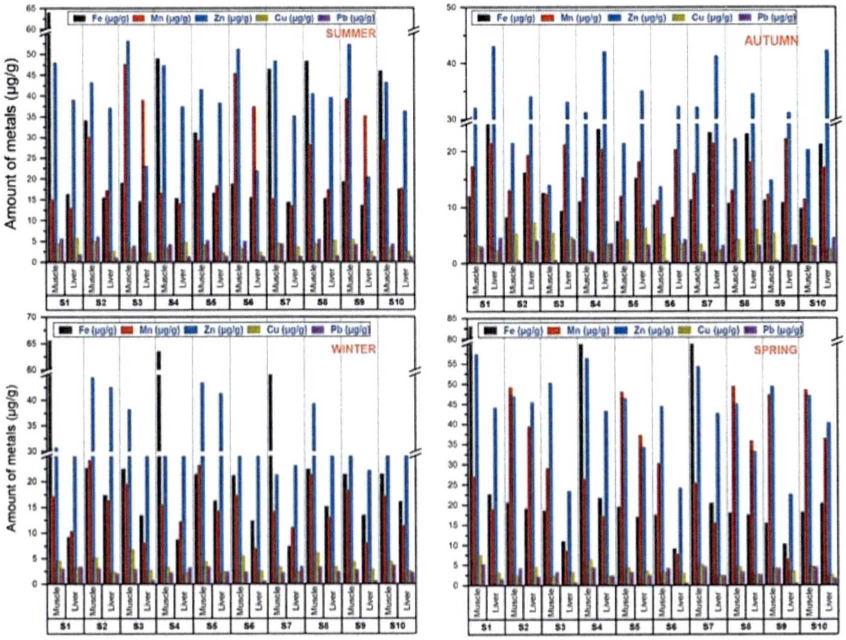

Fig. 3.3 Concentration of HMs in Cyprinus carpio fish muscles and liver (µg/g dry wt) at different sites along Kunhar River stations

adsorption, such as that accomplished by Metallothionein's, significantly contributes to the high concentrations of different HMs in the liver. High levels of metals in the liver were found to have the same effects studied by (Tepe et al. 2008).

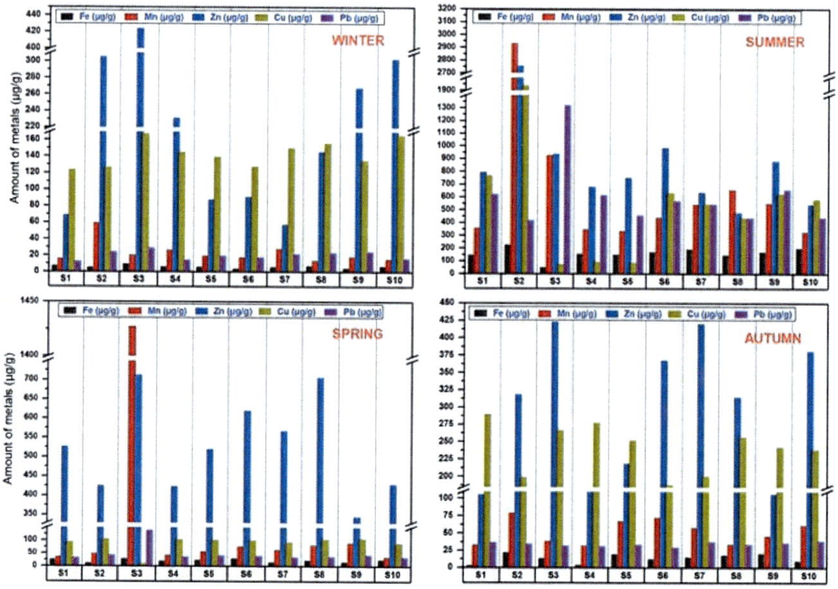

Fig. 3.4 BAF of various HMs in Cyprinus carpio fish flesh of Kunhar River at a different stations

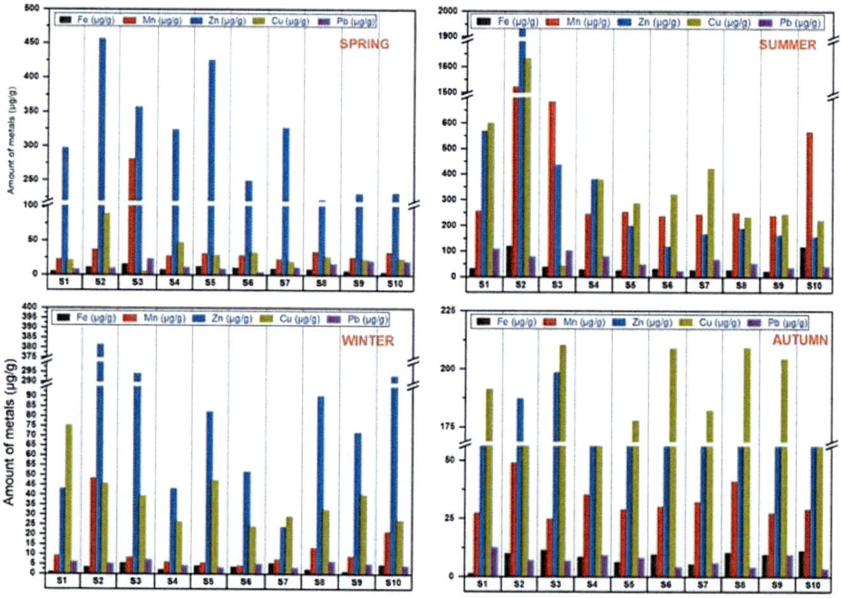

Fig. 3.5 BAF of various HMs in the Cyprinus carpio fish liver of Kunhar River at a different stations

3.5 Conclusion

The research related to the water quality and fish in ten different sites along the Kunhar River, Pakistan. Fish species "Cyprinus carpio" has been used as an indicator to examine metal pollution by measuring their concentration in flesh and liver. Management of high domains around the Kunhar River is, definitely, a massive task for the global community, due to the limited active periods for fieldwork entanglements in the Himalayan region. The amounts of metals currently found in the water need not cause any type of threat to humans or aquatic life. The impacts of climate change on river habitats all around the world are now obvious and are just expected to worsen. We anticipate that our results can inform future predictions of whether climate change will affect water supply and demand. Due to growing demands from tourist economy, the land is increasingly strained, which increases the risk of soil degradation, pollution, damage to ecosystems, and even protection from already threatened species. Historically river species have been primarily linked to climate change; thus, it is essential to monitor and maintain environmental resources to make sure that rivers never become dead. Extra focus in the future must be devoted to the various metal types in the context of climate change to better comprehend the possible threat that these metals cause to aquatic species.

3.6 Data Availability Statement

The datasets are available from the corresponding author on reasonable request.

3.7 Competing Interest

The authors declare no competing interests.

3.8 Informed Consent

This article does not contain any studies with human participants performed by any of the authors.

Acknowledgements The resources used in this paper are provided by College of Hydraulic and Environmental Engineering, China Three Gorges University.

Authors Contribution • **Shan-e-hyder Soomro, Xiaotao Shi, and Jiali Guo**—Conceptualization, Methodology.

• **Xiaotao Shi and Jiali Guo**—Funding Acquisition, Project administration, Resources, Supervision.
• **Shan-e-hyder Soomro, Ao li, Caihong Hu**—Writing-original draft preparation, Writing-review and editing.
• **Shan-e-hyder Soomro, Yanqin Bai, Yuanyang Wang, Zhen yao, Kang Rui**—Literature search.
• **Shaista Jalbani**—Data curation, Analysis of data.

Funding This work was supported by the National Natural Science Foundation of China (51922065, 52279069), Projects of National Natural Science Foundation of China (52179018).

References

Ahmed M, Ahmad T, Liaquat M, Abbasi KS, Farid IBA, Jahangir M (2016) Tissue specific metal characterization of selected fish species in Pakistan. Environ Monit Assess 188:1–9
Ai L, Ma B, Shao S, Zhang L (2022) Heavy metals in Chinese freshwater fish: levels, regional distribution, sources and health risk assessment. Sci The Total Environ: 158455
Amarasinghe S (2020) Health risk assessrnent of heavy metals in drinking water: a case study in Lenabatuwa division of southern Sri Lank
Burkhard LP (2021) Evaluation of published bioconcentration factor (BCF) and bioaccumulation factor (BAF) data for per-and polyfluoroalkyl substances across aquatic species. Environ Toxicol Chem 40:1530–1543
Gowhar SA (2018) Global water quality, statistics and its lethal effects on health an overview. Int J Curr Res Life Sci 7:986–994
Guo J, Shi X, Ke S, Li Y, Hu C, Zwain HM, Gu J, Chunyun Z, Li A, Shenghong LJPJOES (2023) Climate change critique on dams and anthropogenic impact to mediterranean mountains for freshwater ecosystem—a review:32
Hillebrand H, Kahlert M (2002) Effect of grazing and water column nutrient supply on biomass and nutrient content of sediment microalgae. Aquat Bot 72:143–159
Hossain M, Patra PK (2020) Water pollution index—a new integrated approach to rank water quality. Ecol Ind 117:106668
Huang H, Li Y, Zheng X, Wang Z, Wang Z, Cheng X (2022) Nutritional value and bioaccumulation of heavy metals in nine commercial fish species from Dachen Fishing Ground, East China Sea. Scientific Reports 12:6927
Isa BK, Amina SB, Aminu U, Sabo Y (2015) Health risk assessment of heavy metals in water, air, soil and fish. Afr J Pure Appl Chem 9:204–210
Jia Y, Wang L, Qu Z, Wang C, Yang Z (2017) Effects on heavy metal accumulation in freshwater fishes: species, tissues, and sizes. Environ Sci Pollut Res 24:9379–9386
Li C, Wang H, Liao X, Xiao R, Liu K, Bai J, Li B, He Q (2022) Heavy metal pollution in coastal wetlands: a systematic review of studies globally over the past three decades. J Hazard Mater 424:127312
Liu M, Xu Y, Nawab J, Rahman Z, Khan S, Idress M, Ali A, Ahmad R, Khan SA, Khan A (2020) Contamination features, geo-accumulation, enrichments and human health risks of toxic heavy metal (loids) from fish consumption collected along Swat river, Pakistan. Environ Technol Innovation 17:100554
Mamdouh S, Mohamed AS, Mohamed HA, Fahmy WS (2022) The effect of zinc concentration on physiological, immunological, and histological changes in crayfish (Procambarus clarkii) as bio-indicator for environment quality criteria. Biol Trace Elem Res 200:375–384
Mertler CA, Vannatta RA, Lavenia KN (2021) Advanced and multivariate statistical methods: practical application and interpretation. Routledge

Mitra S, Chakraborty AJ, Tareq AM, Emran TB, Nainu F, Khusro A, Idris AM, Khandaker MU, Osman H, Alhumaydhi FA (2022) Impact of heavy metals on the environment and human health: Novel therapeutic insights to counter the toxicity. J King Saud Univ-Sci:101865

Patel V, Parikh P (2013) Assessment of seasonal variation in water quality of River Mini, at Sindhrot, Vadodara. Int J Environ Sci 3:1424–1436

Qadir A, Malik RN (2011) Heavy metals in eight edible fish species from two polluted tributaries (Aik and Palkhu) of the River Chenab, Pakistan. Biol Trace Elem Res 143:1524–1540

Rajeshkumar S, Li X (2018) Bioaccumulation of heavy metals in fish species from the Meiliang Bay, Taihu Lake, China. Toxicol Rep 5:288–295

Raman RK, Talukder A, Mahanty A, Sarkar DJ, Das BK, Bhowmick S, Samanta S, Manna SK, Mohanty BP (2022) Arsenic bioaccumulation and identification of low-arsenic-accumulating food fishes for aquaculture in arsenic-contaminated ponds and associated aquatic ecosystems. Biol Trace Elem Res 200:2923–2936

Saddique U, Muhammad S, Tariq M, Zhang H, Arif M, Jadoon IA, Khattak NU (2018) Potentially toxic elements in soil of the Khyber Pakhtunkhwa province and Tribal areas, Pakistan: evaluation for human and ecological risk assessment. Environ Geochem Health 40:2177–2190

Sattari M, Bibak M, Bakhshalizadeh S, Forouhar Vajargah M (2020) Element accumulations in liver and kidney tissues of some bony fish species in the Southwest Caspian Sea. J Cell Mol Res 12:33–40

SHAHI S (2022) Water pollution: perceptions, source and variety of factors. Social Sci J Adv Res 2:12–17

Soomro S-E-H, Hu C, Boota MW, Ahmed Z, Chengshuai L, Zhenyue H, Xiang L, Soomro MHAA (2022) River Flood susceptibility and basin maturity analyzed using a coupled approach of geo-morphometric parameters and SWAT model. Water Resour Manage 36:2131–2160

Soomro S-E-H, Hu C, Boota MW, Wu Q, Soomro MHAA, Zhang L (2021) Assessment of the climatic variability of the Kunhar River Basin, Pakistan. Water 13:1740

Soomro S-E-H, Shi X, Guo J, Hu C, Zwain HM, Jalbani S, Li Y, Guo Y, Chunyun Z, Gu JJJOW, Change C (2023a) Anthropocentric perspective on climatic variability, potentially toxic elements, and health risk assessment in the Mansehra district: a case study of the Kunhar River. Pakistan 14:1132–1146

Soomro S-E-H, Shi X, Guo J, Hu C, Zwain HM, Liu C, Khan MZ, Niu C, Zhao C, Ahmed Z (2023b) Appraisal of climate change and source of heavy metals, sediments in water of the Kunhar River watershed, Pakistan. Nat Hazard: 1–19

Soomro S-E-H, Shi X, Guo J, Ke S, Hu C, Asad M, Jalbani S., Zwain HM, Khan P, Boot MWJAWS (2023c) Are global influences of cascade dams affecting river water temperature and fish ecology? 13:106

Tepe Y, Türkmen M, Türkmen A (2008) Assessment of heavy metals in two commercial fish species of four Turkish seas. Environ Monit Assess 146:277–284

Ustaoğlu F, Islam MS, Tokatli C (2022) Ecological and probabilistic human health hazard assessment of heavy metals in Sera Lake Nature Park sediments (Trabzon, Turkey). Arab J Geosci 15:597

Vinodhini R, Narayanan M (2008) Bioaccumulation of heavy metals in organs of fresh water fish Cyprinus carpio (Common carp). Int J Environ Sci Technol 5:179–182

Chapter 4
Moroccan Estuary Water Management: Strategies for Development and Sustainability

S. Haddout, K. L. Priya, Joan Cecilia C. Casila, A. M. Hoguane, and I. Ljubenkov

Abstract Moroccan estuaries play a vital role in maintaining the ecological balance of the nation, acting as dynamic connectors between freshwater rivers and coastal saline environments. The effective management of these estuarine water resources is essential for ensuring sustainable development in the region. Striking a delicate balance between developmental goals and long-term environmental sustainability is crucial, given the pivotal role estuaries play in supporting diverse ecosystems, fostering livelihoods, and securing water resources. The significance of these estuaries lies in their ability to act as transitional zones, facilitating the exchange of nutrients, sediments, and organic matter between freshwater and marine environments. This ecological interplay sustains a rich biodiversity and contributes to the overall health of coastal ecosystems. Moreover, estuaries serve as crucial habitats for various species, including commercially important fish, making them integral for the fisheries sector and the livelihoods of local communities. As Morocco pursues its developmental objectives, it becomes imperative to adopt a comprehensive approach to the management of estuarine waters. This involves implementing strategies that not only support economic and infrastructural development but also prioritize the preservation of environmental integrity. Sustainable practices, such as responsible land use planning, pollution control measures, and the protection of critical habitats, must be

S. Haddout (✉)
Faculty of Science, Department of Physics, Ibn Tofail University, B.P. 133 Kenitra, Morocco
e-mail: soufian.haddout@gmail.com

K. L. Priya
Department of Civil Engineering, TKM College of Engineering, Kollam 691005, India

J. C. C. Casila
Land and Water Resources Division, IABE, CEAT, University of the Philippines Los Baños, College, Laguna 4031, Philippines

A. M. Hoguane
Centre for Marine Research and Technology, Eduardo Mondlane University, PO Box 128, Quelimane, Mozambique

I. Ljubenkov
Water Development Ltd., Kvaternikova 7, Split, Croatia

Z. Wan (ed.), *Water Resources Management and Water Pollution Control*, Environmental Science and Engineering, https://doi.org/10.1007/978-3-031-53456-0_4

integrated into the management framework. This study aims to contribute valuable insights and recommendations to support the development of effective strategies that balance the ecological integrity of estuaries with the socio-economic development goals of the nation.

Keywords Morocco estuary · Management · Development · Sustainability

4.1 Introduction

Moroccan estuaries play a crucial role in the country's ecological and economic landscape, serving as dynamic interfaces between freshwater rivers and the saline waters of the ocean (Haddout et al. 2019; Urama and Ozor 2010). These estuarine ecosystems are essential for sustaining biodiversity, providing habitats for numerous species, and supporting various socio-economic activities. As Morocco faces increasing challenges related to water scarcity, pollution, and climate change, effective estuary water management becomes imperative for both ecological preservation and sustainable development (Combe 1966; Haddout 2020). This study focuses on exploring strategies for the development and sustainability of Moroccan estuary water management. The delicate balance between environmental conservation and meeting the growing demands of a developing nation poses a unique set of challenges that require innovative and adaptive approaches. By examining current issues, understanding the ecological significance of estuaries, and identifying potential solutions, this research aims to contribute to the formulation of comprehensive and forward-thinking water management strategies. The management of Moroccan estuaries necessitates a multidisciplinary approach, involving hydrology, ecology, engineering, and policy development (Rhomad et al. 2023; Haddout and Maslouhi 2018). As the nation pursues economic growth and urbanization, the pressure on estuarine ecosystems intensifies. Urban expansion, industrial activities, and agricultural practices can all contribute to habitat degradation, water quality deterioration, and disruption of the delicate balance that sustains these vital ecosystems. This study seeks to address key questions, such as how to balance the needs of agriculture, industry, and urban centers with the imperative to conserve estuarine ecosystems. It will explore the potential of innovative technologies, community-based conservation initiatives, and policy frameworks to ensure the sustainable use of estuary resources. Additionally, a focus on community engagement and collaboration is crucial for the success of any water management strategy. Understanding and incorporating local knowledge, values, and needs into the decision-making process can enhance the resilience of estuarine ecosystems and foster a sense of ownership among the communities that depend on them. The sustainable management of Moroccan estuary waters is a complex and pressing challenge that requires a holistic and collaborative approach (Blidi and Fekhaoui 2003; Priya et al. 2022). This study aims to contribute valuable insights and recommendations to support the development of effective strategies that balance the ecological integrity of estuaries with the socio-economic development goals of the nation.

4.2 Moroccan Estuaries

The Moroccan estuaries (Fig. 4.1) are characterized by dynamic environmental conditions and are home to diverse flora and fauna. Here are some key points about Moroccan estuaries:

Biodiversity:

Estuaries in Morocco support a rich biodiversity, including various species of fish, crustaceans, and birds. The mixing of freshwater and saltwater creates a unique habitat that supports the life cycles of many marine organisms.

Economic Importance:

The estuarine areas are often important for local economies, providing habitats for commercially important fish and shellfish species. Fishing communities rely on these estuaries for their livelihoods, as they serve as breeding and nursery grounds for many marine species.

Environmental Challenges:

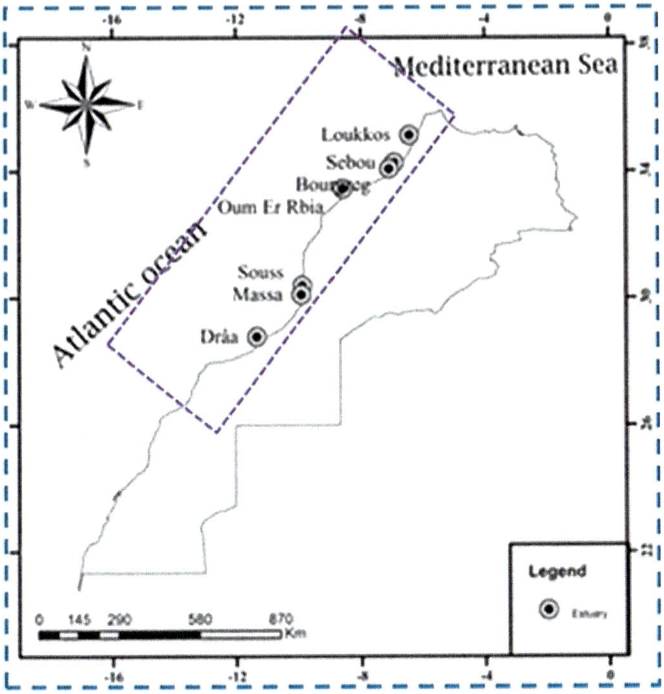

Fig. 4.1 Moroccan estuaries

Moroccan estuaries, like many worldwide, face environmental challenges such as pollution, habitat degradation, and overexploitation of resources. Human activities, industrial discharges, and agricultural runoff can introduce pollutants into these sensitive ecosystems.

Conservation Efforts:

Recognizing the ecological importance of estuaries, there are conservation efforts in place to protect and manage these areas sustainably. These efforts may include the establishment of marine protected areas, regulations on fishing practices, and initiatives to reduce pollution.

Tourism and Recreation:

Some estuarine areas in Morocco may also attract tourists and recreational activities. Visitors may be drawn to the unique landscapes, bird watching opportunities, and the chance to experience the cultural and natural richness of these coastal zones.

Climate Change Impact:

Estuaries are vulnerable to the impacts of climate change, including rising sea levels and changes in temperature and precipitation patterns. These changes can affect the ecology of estuarine ecosystems and the communities that depend on them.

Some specific examples of estuaries in Morocco include those along the Atlantic and Mediterranean coasts, where rivers like the Sebou and Oued Loukos discharge into the sea (Haddout et al. 2016, 2017). These estuaries contribute to the overall environmental health and economic well-being of the regions they are located in.

4.3 Development and Sustainability

The development and sustainability of Moroccan estuaries are crucial for the ecological health of these unique ecosystems and the well-being of the surrounding communities. In the case of Moroccan estuaries, several key factors contribute to their development and sustainability:

Environmental Protection and Conservation:

- Implement and enforce strict environmental regulations to protect estuarine ecosystems from pollution, habitat destruction, and over-exploitation of natural resources.
- Establish and maintain protected areas within estuaries to preserve critical habitats for migratory species, such as fish and birds.

Integrated Coastal Zone Management (ICZM):

- Develop and implement ICZM plans that take into account the dynamic interactions between land, water, and coastal ecosystems.
- Promote sustainable land use practices to prevent soil erosion and maintain water quality in estuarine areas.

Community Engagement and Stakeholder Collaboration:

- Involve local communities, fisher folk, and other stakeholders in decision-making processes related to estuary management.
- Raise awareness about the importance of estuaries and promote community-based initiatives for conservation and sustainable resource use.

Water Quality Management:

- Monitor and control industrial discharges, agricultural runoff, and other sources of pollution that can degrade water quality in estuaries.
- Implement measures to reduce sedimentation and nutrient loading, which can harm the balance of estuarine ecosystems.

Climate Change Adaptation:

- Develop strategies to address the impacts of climate change on estuarine ecosystems, such as sea-level rise, changes in temperature and precipitation patterns, and increased frequency of extreme weather events.
- Consider nature-based solutions, such as restoring mangroves and other coastal vegetation, to enhance the resilience of estuarine ecosystems to climate change.

Infrastructure Planning:

- Plan and construct infrastructure, such as ports and urban developments, with careful consideration for the impact on estuarine ecosystems.
- Implement measures to minimize habitat destruction and maintain natural water flow patterns.

Scientific Research and Monitoring:

- Conduct regular scientific research and monitoring programs to assess the health of estuarine ecosystems.
- Use data-driven approaches to inform decision-making and identify emerging threats to sustainability.

Education and Capacity Building:

- Provide educational programs to schools and communities to foster a better understanding of estuarine ecosystems and the importance of sustainable practices.
- Build the capacity of local institutions to manage and monitor estuarine resources effectively.

The sustainable development of Moroccan estuaries requires a holistic and collaborative approach, involving government agencies, local communities, non-governmental organizations, and the private sector. By integrating environmental protection, community engagement, and climate resilience, Morocco can ensure the long-term health and sustainability of its estuarine ecosystems.

4.4 Conclusions

In conclusion, the development and sustainability of Moroccan estuaries are imperative for the ecological balance of these vital ecosystems and the well-being of local communities. A comprehensive and integrated approach is essential, encompassing environmental protection, community engagement, climate change adaptation, and careful infrastructure planning. By implementing and enforcing environmental regulations, promoting sustainable practices, and involving local stakeholders in decision-making processes, Morocco can safeguard its estuarine ecosystems.

Climate change adaptation strategies and ongoing scientific research are crucial components for ensuring the resilience of estuaries in the face of environmental challenges. Additionally, education and capacity building play pivotal roles in raising awareness and empowering communities to actively participate in the conservation and sustainable use of estuarine resources. Ultimately, the sustainable development of Moroccan estuaries requires a commitment to balance economic growth with environmental conservation. By prioritizing the preservation of these unique ecosystems, Morocco can secure a healthier environment, thriving biodiversity, and sustainable livelihoods for present and future generations.

References

Haddout S, Baimik I, Maslouhi A, Igouzal M, Magrane B, Marah H (2019) The influence of spring and neap tide on salt intrusion and stratification in Sebou estuary (Morocco). Int J River Basin Manage 17(1):131–142

Urama KC, Ozor N (2010) Impacts of climate change on water resources in Africa: the role of adaptation. Afr Technol Policy Stud Netw **29**(1):1–29

Combe M (1966) Etude des marées dans l'oued Sebou et des pollutions qu'elles provoquent à l'étiage (Study of tidal cycle in the Sebou estuary during low water). Rapport inédit, Rabat, MTPC/DH DRE, p 108

Haddout S (2020) A power-law multivariable regression equation for salt intrusion length in the Bouregreg estuary, Morocco. Mar Georesour Geotechnol 38(4):417–422

Rhomad H, Khalil K, Elkalay K (2023) Trends and hot spots of coastal science in Moroccan Atlantic coast: a bibliometric analysis. Environ Dev Sustain 1–24

Haddout S, Maslouhi A (2018) One-dimensional hydraulic analysis of the effect of sea level rise on salinity intrusion in the Sebou estuary, Morocco. Mar Geodesy 41(3):270–288

El Blidi S, Fekhaoui M (2003) Hydrologie et dynamique marégraphique de l'estuaire du Sebou (Gharb, Maroc). Bulletin de l'institut Scientifique, Rabat, section Sciences de la Vie 25:57–65

Priya KLA, Haddout S, Casila JCC (2022) Implications of turbulent shear on clay floc break-up along the Atlantic estuary (Bouregreg), Morocco. Int J Sedim Res 37(2):248–257

Haddout S, Igouzal M, Maslouhi A (2016) Analytical and numerical study of the salinity intrusion in the Sebou river estuary (Morocco)—effect of the "Super Blood Moon" (total lunar eclipse) of 2015. Hydrol Earth Syst Sci 20(9):3923–3945

Haddout S, Igouzal M, Maslouhi A (2017) Seawater intrusion in semi-closed convergent estuaries (case study of Moroccan Atlantic Estuaries): application of salinity analytical models. Mar Geodesy 40(5):275–296

Chapter 5
Research Progress on the Removal Effect and Mechanism of Phosphorus from Water by Biomass Materials

Luyi Nan, Yuting Zhang, Min Liu, Yuxuan Zhu, and Liangyuan Zhao

Abstract Eutrophication of water body is one of the major environmental pollution problems faced by all countries in the world. This article mainly discusses various biomass materials for removing phosphorus pollution in eutrophication water. Firstly, it introduces the current situation of phosphorus pollution in eutrophication water is introduced, and then various mainstream phosphorus removal technologies are described, such as physical method, chemical method and biological method, especially introduces the biomass phosphorus removal technology in adsorption method, which is cost-effectively and efficient. All kinds of biomass materials have been listed, such as biomass-based phosphorus removal materials, biochar phosphorus removal materials and modified biomass phosphorus removal materials, and their application range, advantages and disadvantages as well as their preparation and modification methods are introduced. In addition, different purification mechanisms such as ion exchange, physicochemical adsorption, coordination adsorption, surface precipitation and hydrogen bonding are also introduced. Finally, the current challenges and future development directions of phosphorus removal from biomass materials are discussed and prospected.

Keywords Eutrophication · Phosphorus removal technology · Biomass

---

L. Nan · Y. Zhang · M. Liu · Y. Zhu · L. Zhao (✉)
Basin Water Environmental Research Department, Changjiang River Scientific Research Institute, Wuhan 430010, China
e-mail: zhaoliangyuannew@163.com

Key Lab of Basin Water Resource and Eco-Environmental Science in Hubei Province, Wuhan 430010, China

Innovation Team for Basin Water Environmental Protection and Governance of Changjiang Water Resources Commission, Wuhan 430010, China

Z. Wan (ed.), *Water Resources Management and Water Pollution Control*, Environmental Science and Engineering, https://doi.org/10.1007/978-3-031-53456-0_5

5.1 Introduction

Eutrophication of water is one of the major environmental pollution problems (Li et al. 2006). The excess of nutrients such as nitrogen and phosphorus in water leads to the outbreak of algae growth in the water, which not only causes many problems to the ecosystem, but also may pose a great threat to the quality of water and the health of life (Le et al. 2010; Hou et al. 2018; Singh et al. 2018). Since 2016, phosphorus has become the main pollution in the Yangtze River (Liu et al. 2020).

How to efficiently manage the eutrophication of water is the research hotspot in related fields. Currently, the main methods used to remove phosphorus from eutrophic water bodies are physical methods, chemical methods, biological methods, and so on (Li 2014, 2021). Physical methods include blowing off method, ionization method, and other methods (Yuan et al. 2005). Chemical methods include precipitation, electrolysis and catalytic oxidation (Wu et al. 2022). Biological methods include microbial remediation and phytoremediation (Jing et al. 2017; Ru 2017). Among these methods, the adsorption method is considered to be more suitable for treating phosphorus wastewater because of its wide applicability, simple treatment process, stable effect, and no secondary pollution, etc. (Liu 2018). Cost-effectively and high-efficiency physical adsorbents have attracted great interest, and the main commercial adsorbents are zeolites, bentonite, and biochar. Among them, biomass phosphorus removal materials have been widely studied by researchers because of their relatively simple production process, good phosphorus removal effect, environmental-friendly, and the raw material is widely sourced and renewable and easily degradable.

5.2 Application of Natural Biomass Materials for Phosphorus Removal

Biomass refers to the use of the atmosphere, water, soil, etc., through photosynthesis to form a variety of organisms, including all plants, microorganisms, animals and their wastes that feed on plant microorganisms. At present, biomass adsorption materials used for phosphorus removal from water bodies mainly include biomass-based phosphorus removal materials, biochar phosphorus removal material, modified biomass phosphorus removal material, etc.

5.2.1 Biomass-Based Phosphorus Removal Materials

Biomass-based phosphorus removal materials utilize the unique chemical structure and functional properties of biomass in the water treatment process to adsorb pollutants through solid–liquid separation. Common biomass-based phosphorus removal materials can be obtained from multiple sources, including microorganisms such as algae, bacteria, and fungi (Li 2022), natural polysaccharides derived from living organisms such as cellulose, chitosan, and sodium alginate (Yu et al. 2021; Hamedi et al. 2018; Yi et al. 2018), and agricultural and forestry wastes such as barks, straw, and fruit peels (Li 2015). Compared with organic polymer-based phosphorus removal materials, biomass-based phosphorus removal materials can be biodegradable and renewable, environmentally friendly, and more in line with the green development concept of "carbon neutrality and carbon compliance" (Zhang et al. 2019).

Waste agricultural biomass-based phosphorus removal materials are more practical than other kinds of biomass-based phosphorus removal materials, unprocessed cornstalks straw, rice straw, wheat straw, reed, woods, and other waste agricultural biomass-based phosphorus removal materials contain a number of surface functional groups, which can enhance the adsorption of pollutant ions. For instance, rice straw, luffa and corn cobs were added into the eco-bag for experiments, and the nitrogen removal rates were 78.4%, 77.8% and 73.6%, respectively, and the phosphorus removal rates could reach 65.1%, 62.2% and 63.1%, respectively (Wang et al. 2021).

5.2.2 *Biochar Phosphorus Removal Materials*

Biochar phosphorus removal materials have high porosity and specific surface area, which can effectively adsorb harmful substances in water, such as heavy metal ions, organic matter, nitrogen, phosphorus and so on. It is mainly composed of two parts: one is carbonized biomass, such as wood, bamboo, coconut shell and rice husk; the other is activated biochar.

Biochar phosphorus removal material has the following characteristics: (1) Strong adsorption capacity: it has high porosity and specific surface area, which can better adsorb the target substances. (2) Good stability: it is characterized by good chemical stability, and its adsorption effect will not be affected by changes in environmental temperature, humidity and other factors. (3) Wide applicability: it is suitable for treating different kinds of water pollutants.

It has been reported that biochar prepared from the leaves of the coneflower tree harvested in the manganese mining area for the removal of phosphate from water, and it had good adsorption performance under neutral and alkaline conditions, at the same time it was better simulated by the Langmuir model at temperatures of up to 35 °C, with a theoretical maximum adsorption capacity of 28.2 mg/g (Qiu 2020). Ma et al. used the equilibrium adsorption method to study the phosphorus adsorption characteristics of cattle dung biochar. The results showed that the optimal initial pH value of biochar phosphorus removal material was 7.0; when the dose amount was 0.1 g, the phosphorus removal rate was the highest, and the theoretical maximum adsorption amount of Langmuir model was 4.7094 mg/g (Ma et al. 2015).

The preparation process of biochar phosphorus removal material mainly includes steps such as biomass carbonization, activation treatment and sieving. And biomass carbonization is to crack biomass at high temperature and transform it into charcoal material; activation treatment is to inject gas or chemical reagents into the surface of biochar to increase its porosity and specific surface area and improve its adsorption performance; and screening is to classify the biochar and remove particles that do not meet the specifications (Peng 2023).

In addition the preparation methods of biochar phosphorus removal materials mainly include high-temperature pyrolysis and hydrothermal carbonization.

1. **High-Temperature pyrolysis**

High-temperature Pyrolysis (Pyrolysis) is a method of preparing biochar by controlling pyrolysis of biomass feedstocks in an oxygen-poor or very low-oxygen environment (Sun 2019). For example, in the coconut shell biochar preparation, when the pyrolysis temperatures were 500 °C and 700 °C, the pH was significantly higher than the pH at 300 °C, and the phosphorus removal rate increased effectively (Zhong et al. 2023).

2. **Hydrothermal carbonization**

Hydrothermal carbonization technology is one of the main technologies with the most development potential and prospect for the efficient resource utilization of biomass. Hydrothermal carbonization is a mild thermochemical reaction carried out in a closed system with water as solvent at a certain temperature, reaction time and autogenous pressure (Wang et al. 2019; Li et al. 2023). The biochar produced by hydrothermal carbonization has high yield, rich in surface oxygen-containing functional groups, high porosity, and good hydrothermal stability, which has the potential to be applied in the field of adsorption (Fang et al. 2018).

When the residual sludge from an urban wastewater treatment plant was used as the raw material, hydrothermal carbon was prepared by hydrothermal carbonization method at different temperatures and times for adsorption of Cr(VI) in water, the removal rate of Cr(VI) could reach 99.81% for the wastewater with an initial mass concentration of 50 mg/L at the temperature of 25°C, pH 2.5, and the dosage of hydrothermal carbon of 10 g/L (Fang et al. 2021).

5.2.3 Modified Biomass Phosphorus Removal Materials

The limited adsorption capacity of common biomass materials when used as adsorbents results in generally low adsorption efficiencies, so in order to improve the adsorption performance of biochar for better removal of pollutants, more and more studies are tending to modify the biochar (Rivera-Utrilla et al. 2011). The Modified biochar has the advantages of larger surface area and increasing numbers of functional groups, so the modified biochar can significantly improve the adsorption capacity.

Modified phosphorus removal materials mainly include acid modified biochar and alkali modified biochar, metal-modified biochar, oxidizer activated biochar.

Acid-modified and alkaline-modified biochar is prepared by soaking biochar in acid or alkaline solution and then washing and drying it to produce modified biochar, the advantage of it is that it can enhance the adsorption capacity of the original biochar, so that the specific surface area and pore structure of the object charcoal increase, and the content and type of functional groups rise. Using the balsam harvested in the wetland as the raw material, the phosphorus removing effect of the acid-modified biochar prepared after acid-modification was well compared with that of the original biochar, which could be increased from 65 to 94% of T7 (Wang 2022).

Metal-modified biochar includes iron modification, magnesium modification, lanthanum modification, composite metal modification, etc., and its modification is to load metal onto the surface of biochar to improve the adsorption capacity of the biochar. Using forest waste as raw material, magnesium-modified biochar was prepared through dry pyrolysis and wet pyrolysis. Under the optimal conditions, the adsorption capacity of Mg-modified biochar for nitrogen and phosphorus was 35.28, 110.29 mg/g, respectively (Zhu et al. 2018).

Oxidizer activation of biochar such as modification using hydrogen peroxide increases the number of oxygen-containing functional groups on the surface of the biochar and enriches the internal pore structure of the biochar.

5.3 Water Purification Mechanism of Biomass Phosphorus Removal Materials

5.3.1 Ion Exchange

The ion exchange process is the process of mass transfer (including external diffusion and internal diffusion) and chemical reaction (ion exchange reaction) between liquid and solid phases. When phosphate ions form ionic bonds with cations on the surface of a solid, some anions that were previously attached to the metal ion surface (Sparks 2001). Because ion exchange is reversible and highly efficient, it is the main reaction route in the adsorption process, and it has great potential for phosphorus recovery.

5.3.2 Physico-Chemical Adsorption

Physical adsorption is a phenomenon formed by molecules forming an adsorption layer on a surface. Due to the large specific surface area and pore structure of biochar, it can provide many adsorption sites so that dissolved pollutants in water can quickly attach to the surface of biochar.

Chemical adsorption refers to the formation of some chemical bonding between pollutants and adsorbent materials through chemical reactions such as covalent bonding and coordination bonding. The functional groups in biochar, such as hydroxyl and carboxyl groups, can react with organic matter in water by hydrogen bonding, ionic bonding and other reactions, so as to remove organic matter, heavy metals and other pollutants in water. In addition, there are some functional groups with cation exchange capacity in biochar, which can effectively remove nutrients such as ammonia, nitrogen and phosphorus from water.

5.3.3 Coordination Adsorption

Ionic coordination is a general physical cross-linking by the ligands of the polymer chain, which can form a strong complex with metal ions in the aqueous phase. During ligand exchange, adsorbed anions, such as phosphates can form coordination bonds with metal cations on the surface of the adsorbent, resulting in the release of weakly coordinating anions or molecules. Therefore, the phosphate forms an internal complex with the metal cation on the surface of the adsorbent and is adsorbed onto the differently charged surface to create a negative charge, shifting the zero charge point to a lower pH.

5.3.4 Surface Precipitation

Precipitation of phosphate may occur when the ionic product of the phosphate precipitate in solution exceeds its solubility product constant. According to the thermodynamic solubility product principle, surface deposition of metal phosphates occurs even at solution concentrations of phosphate and metal ions that are lower than expected, and metal precipitates in the solution phase (Sparks 2001).

5.3.5 Hydrogen Bonding

The formation of hydrogen bonds is due to the interaction force formed between partially positively charged hydrogen atoms in polar molecules and negatively charged atoms (such as oxygen, nitrogen, etc.) (Li et al. 2016). The mechanism of hydrogen bond adsorption is the formation of hydrogen bonds on the adsorbent surface to separate the target molecules from the mixture.

5.4 Conclusion

This article summarizes the mainstream natural biomass adsorbent materials: biomass-based phosphorus removal materials, biochar phosphorus removal materials and modified biochar phosphorus removal materials, and their advantages and disadvantages as well as preparation and modification methods are described respectively. The mechanism of phosphorus adsorption of biochar materials has been also introduced.

Among these materials, modified biochar phosphorus removal materials are the most promising of the current natural biomass phosphorus removal materials. Therefore, the modified biochar phosphorus removal materials need to be explored more deeply in the subsequent research.

Most natural water bodies are complex and changeable. In the laboratory environment, only one source of pollution is usually simulated in the laboratory environment, which lacks authenticity. In addition, the preparation method of biochar still cannot meet the needs of environmental protection. Therefore, materials should be made to make more environmentally friendly adsorption materials, such as biomass aerogel.

Acknowledgements Acknowledgements: Funding

This work was financially supported by National Key Research and Development Program of china (No.2023YFC3208702) and the Fundamental Research Funds for the Central Public Welfare Research Institutes (No. CKSF2021743/HL, No. CKSF2023311/HL), and the State-level Public Welfare Scientific Research Institutes Basic Scientific Research Business Project of China (No. CKSF2023337/SH).

References

Fang J, Zhan L, Ok SY, et al (2018) Minireview of potential applications of hydrochar derived from hydrothermal carbonization of biomass. J Ind Eng Chem 57:15–21

Fang J, Li Y, Tang Q, et al (2021) Preparation and characterization of sludge hydrothermal carbon for adsorption of Cr(VI) in wastewater. Water Treat Technol 47(9):52–57

Hamedi H, Moradi S, Hudson SM et al (2018) Chitosan based hydrogels and their applications for drug delivery in wound dressings: a review. Carbohyd Polym 199:445–460

Hou Q, Meng P, Pei H et al (2018) Phosphorus adsorption characteristics of alum sludge: adsorption capacity and the forms of phosphorus retained in alum sludge. Mater Lett 229:31–35

Jing D, Qing-Liang Z, Jun Z, et al (2017) Hybrid electrooxidation and adsorption process for the removal of ammonia in low concentration chloride waste water. Environ Sci Pollut Res Int 24(6):5098–5105

Le C, Zha Y, Li Y, et al (2010) Eutrophication of lake waters in China: Cost, causes, and control. Environ Manage 45(4):662–668

Li J (2015) Research on wheat straw charcoal modified activation and its nitrogen and phosphorus adsorption effect. Chinese Academy of Agricultural Sciences

Li J (2022) Research on adsorption properties and application of biomass-based functional materials. University of Electronic Science and Technology

Li S, Do H, Shu J, et al (2006) Discussion on water environment problems and water ecosystem restoration of lakes in China. China Water Resour 13:14–17

Li J, Tang Y, Huang Y, et al (2014) Research progress of biomass adsorption treatment of oil pollution in water bodies. Environ Pollut Prev 36(10):79–87
Li Z, Wang X, Peng S, et al (2021) Research progress on adsorption of heavy metals by chemically modified cellulose. Energy Environ Prot 35(2):1–6
Li X, Fei X, Cao Y (2023) Research progress on the preparation and application of biochar. Chem Technol Dev 52(5):37–40
Li M, Liu J, Xu Y, Qian G (2016) Phosphate adsorption on metal oxides and metal hydroxides: a comparative review. Environ Rev 24(3)
Liu M (2018) Research on purification effect and disposal of corn cob biochar substrate in artificial wetland. Wuhan University
Liu L-S, Huang G-X, Wang F, et al (2020) Main problems, situation and countermeasures of water ecological and environmental security in the Yangtze River basin. Environ Sci Res 33(5):1081–1090
Ma F, Zhao G, Zhong J, et al (2015) Research on the adsorption characteristics of phosphorus by cow dung biochar and its influencing factors. China Environ Sci 35(4):1156–1163
Peng F (2023) Characterization of rhodamine B adsorption in water by rapeseed straw-based biochar. Water Resour Sci Cold Reg Eng 6(3):67–71
Qiu GQ (2020) Adsorption of phosphate in water by structural tree biochar and its effect on soil chemical properties. Central South Forestry University of Science and Technology
Rivera-Utrilla J, Sánchez-Polo M, Gómez-Serrano V, et al (2011) Activated carbon modifications to enhance its water treatment applications: an overview. J Hazard Mater 187(1):1–23
Ru C (2017) Research progress of phosphorus removal technology in eutrophic waters. Guangdong Chem Ind 44(23):100–114
Singh NB, Nagpal G, Agrawal S et al (2018) Water purification by using adsorbents: a review. Environ Technol Innov 11:187–240
Sparks LD (2001) Elucidating the fundamental chemistry of soils: Past and recent achievements and future frontiers. Geoderma 100(3):303–319
Sun X (2019) Research on the preparation and application of biochar. Rural Sci Technol 206(2):115–116
Wang L (2022) Degradation of sulfadiazine by oxygen-doped biochar-activated perovskite and reuse of elemental iron. University of Science and Technology of China
Wang L, Chang Y, Li A (2019) Hydrothermal carbonization for energy-efficient processing of sewage sludge: A review. Renew Sustain Energy Rev 108:423–440
Wang Q, Shi L, Yang X, et al (2021) Effectiveness of waste biomass-enhanced ecological bag for nitrogen and phosphorus removal. J Southeast Univ (Natural Science Edition) 51(1):138–144
Wu Q, Tan M, Chi D (2022) Research progress of nitrogen and phosphorus adsorption by biochar in eutrophic waters. J Shenyang Agric Univ 53(5):620–629
Yi X, Sun F, Han Z et al (2018) Graphene oxide encapsulated polyvinyl alcohol/sodium alginate hydrogel microspheres for Cu (II) and U (VI) removal. Ecotoxicol Environ Saf 158:309–318
Yu J, Wang AC, Zhang M et al (2021) Water treatment via non-membrane inorganic nanoparticles/cellulose composites. Mater Today 50:329–357
Yuan D, Zhang M, Gao S et al (2005) Performance and mechanism of adsorption and purification of phosphorus by several clay minerals and clayey soils. Environ Chem 1:7–11
Zhang T, Li Z, Lu Y et al (2019) Recent progress and future prospects of oil-absorbing materials. Chin J Chem Eng 27(6):1282–1295
Zhong W, Fu D, Qi D, et al (2023) Preparation of biochar and its application in water treatment. Water Treat Technol 49(1):26–30
Zhu T, Lu Z, Liu Y, et al (2018) Effect of preparation conditions of magnesium-modified biochar on its nitrogen and phosphorus removal performance. Environ Eng 36(1):37–41

Chapter 6
Modification of Biochar and Its Removal Mechanism of Phosphorus

Luyi Nan, Yuting Zhang, Min Liu, Yuxuan Zhu, and Liangyuan Zhao

Abstract How to effectively control water eutrophication is the current research hotspot in related fields. It has been found that modified biochar has great performance of adsorption of phosphorous. Therefore, this paper mainly discusses the modified biomass materials to remove phosphorus in eutrophication water. Firstlly, the present situation of phosphorus pollution in eutrophication water is introduced, and then phosphorus removal technology is described, especially the modified biomass phosphorus removal materials. The necessity of modification is demonstrated by comparing the physicochemical properties of unmodified and modified biomass. And all kinds of modified biomass materials discussed in this paper are listed as: acid and alkali modified biomass materials, metal-modified biomass materials, oxidizer activated biomass materials, magnetization modified biomass materials and so on. The meaning, application range, advantages and disadvantages of these modified biomass materials are also introduced in detail. In addition, this paper also introduces the mechanism of water purification of modified biomass materials, and briefly discusses its future development trend of the application of biomass sponge and biomass aerogel.

Keywords Phosphorus removal technology · Biomass · Modified biomass

L. Nan · Y. Zhang · M. Liu · Y. Zhu · L. Zhao (✉)
Basin Water Environmental Research Department, Changjiang River Scientific Research Institute, Wuhan 430010, China
e-mail: zhaoliangyuannew@163.com

Key Lab of Basin Water Resource and Eco-Environmental Science in Hubei Province, Wuhan 430010, China

Innovation Team for Basin Water Environmental Protection and Governance of Changjiang Water Resources Commission, Wuhan 430010, China

Z. Wan (ed.), *Water Resources Management and Water Pollution Control*,
Environmental Science and Engineering, https://doi.org/10.1007/978-3-031-53456-0_6

6.1 Introduction

Currently, eutrophication of water bodies is a major hotspot of global environmental issues. It has been reported that more than 50% of the world's lakes and rivers are affected by eutrophication (Shijie et al. 2006). The excess of nitrogen, phosphorus and other nutrients in the water body leads to the growth of algae outbreaks, which not only brings many problems to the ecosystem, but also produces toxic and harmful gases, and deteriorates the environment and jeopardize people's health.

How to carry out the efficient treatment of the eutrophic water body is also the top priority of the current research. The physical adsorption materials because of their low cost and high efficiency of adsorption have been widely concerned. At present, the commercial adsorbents mainly include bentonite, zeolite, activated carbon and biochar, etc. Among them, biomass phosphorus removal material is environmentally friendly and easy to obtain, which has great advantages compared with other classes of adsorbents.

However, the limited adsorption capacity of unmodified biomass materials leads to a general lack of adsorption efficiency, so in order to improve the adsorption performance of biochar for better removal of pollutants, more and more researches are tending to modify the biochar (Rivera-Utrilla et al. 2011). Modified biochar has the advantages of larger specific surface area and numbers of functional groups, which significantly enhances the adsorption capacity (Xuefei et al. 2018).

Therefore, this paper introduces modified biomass from three aspects: modified biomass materials, preparation of modified biomass and its water purification mechanism.

6.2 Modified Biomass Materials

6.2.1 Physical and Chemical Properties of Modified Biochar

The physicochemical properties of biochar are shown in Table 6.1 shown (Mengmeng 2019), where OC (Original Carbon) is the initial biochar and the rest are the modified biochar with different metals. The molar ratios of H/C and O/C of unmodified and modified biochar were in the range of 0.023–0.048 and 0.008–0.578, and the lower H/C of the six modified biochars indicated the high degree of aromaticity and plant charring, and the biochars contained a large amount of plant organic residue (Yuan et al. 2006). The O/C and (O + N)/C molar ratios of the modified biochar were higher than those of the unmodified biochar, which indicated that there were a large number of polar functional groups in the modified biochar and its surface was more hydrophilic, and polarity is stronger, so it was more conducive to adsorption of pollutants, such as NH_4^+-N, PO_4^{3-}-P (Cui et al. 2016).

Table 6.1 Physical and chemical properties of biochar

Biochar	C (%)	H (%)	N (%)	O (%)	H/C	O/C	(O + H)/C
Mg-OC	54.67	2.05	0.53	8.54	0.038	0.156	0.166
Al-OC	48.1	1.6	0.34	11.54	0.033	0.24	0.247
Fe-OC	34.48	0.81	0.35	19.94	0.023	0.578	0.588
Ca-OC	39.49	1.88	0.26	18.88	0.048	0.478	0.485
La-OC	21.94	0.94	0.2	8.42	0.043	0.384	0.393
OC	77.41	2.36	0.99	6.2	0.031	0.08	0.093

6.2.2 *Different kinds of Modified Biochar*

By modifying biochar in different ways, its surface charge can be transformed so that its adsorption capacity for phosphate can be significantly improved. For example, the surface pores of acid-modified biochar were well improved, and the specific surface area was as high as 434.2 m^2/g, which had a large adsorption potential, and the adsorption capacity could reach 0.97 mg/g. The MgO-modified biochar had a large adsorption capacity of 0.97 mg/g (Ziqiong et al. 2022). The yield of MgO-modified biochar (45.01–56.35%) was higher than that of original biochar (29.29–30.08%), and the adsorption capacities of the original biochar and MgO-modified biochar for phosphorus were 1.88–2.78 mg/g and 28.20–29.22 mg/g, respectively, and the MgO-modified biochar showed an 11-fold greater phosphorus adsorption capacity (Oginni et al. 2020). The specific modified biomass materials are summarized by Table 6.2 (Wenjie et al. 2022) shown, which mainly include acid and alkali modified biomass materials, metal- modified biomass materials, oxidizer activated biomass materials, magnetization modified biomass materials:

6.3 The Preparation of Modified Biomass

The modification preparation methods of biochar mainly include activation method, loaded metal oxide modification and loaded organic modification:

1. **Activation method**

   Activation method refers to the treatment of biochar by chemical or physical methods, which makes its pore structure more developed, with more surface functional groups, and thus improving its adsorption capacity. Commonly used activators include: potassium hydroxide, zinc chloride, phosphoric acid and so on. For example, silkworm sand-based biochar was prepared by impregnation and 400°C impregnation-pyrolysis activation processes using silkworm sand biochar as raw material and KOH as activator for adsorption and removal of cadmium ions (Cd^{2+}) from water bodies. The experimental results showed that the maximum adsorption amounts of the two materials obtained in different ways were 63.80

Table 6.2 Types of modified biomass

Types of modified biomass		Methodologies	Dominance	Bibliography
Acid and alkali modified biomass materials	Acid modification	Soak the biochar in an acidic solution, then wash and dry to obtain modified biochar	Acid-modified biochar can significantly enhance the adsorption capacity of original biochar. After acid modification, the specific surface area and pore structure of the biochar increase, and the content and type of functional groups rise, so that the adsorption capacity of its modified biochar is enhanced	Yanjun et al. (2021)
	Alkali modification	Modified biomass is obtained by soaking the biomass in a lye solution, washing and drying it, or by mixing the lye with the biomass and heating it under oxygen-restricted conditions	Alkali-modified biochar can enhance the adsorption capacity of original biochar, but not as much as metal-modified biochar	Haopu and Bo (2022)
Metal-modified biomass materials	Magnesium-modified biochar	Loading magnesium onto the biomass surface	The presence of calcium and magnesium ions changes the surface chargeability of biochar, increasing the displacement rate of phosphate ions, which in turn improves the phosphorus adsorption efficiency	Yuhong et al. (2022)

(continued)

Table 6.2 (continued)

Types of modified biomass		Methodologies	Dominance	Bibliography
	Iron-modified biochar	Loading iron onto the biomass surface	Iron has a strong attraction with phosphate. Iron oxides are magnetic, and the loading of iron oxides on the surface of biochar significantly increases the phosphorus removal effect of the adsorbents	Xinran et al. (2023)
	Lanthanum-modified biochar	Loading of lanthanum onto the biomass surface	Lanthanum is found in high concentrations in nature and is less expensive than other rare earth elements	Xinfeng et al. (2023)
	Complex metal oxides	Loading the composite metal onto the biomass surface	Complex metal oxides have higher adsorption efficiencies than individual metals due to intermetallic synergies	Jiali (2022)
	Other metal-modified biochar			Sulin and Congyuan (2022)
Oxidizer activated biomass materials		Modification with hydrogen peroxide	The number of oxygen-containing functional groups on the surface increased and the pore structure inside the biochar was enriched	Lulu (2022)
Magnetization modified biomass materials		Iron loading on biochar	The addition of magnetic materials facilitates solid–liquid separation of biomass and facilitates recycling	Lin et al. (2015)

mg/g and 89.15 mg/g, respectively, at the dosage of 0.4 g/L and pH = 5.0 (Yapen et al. 2021).

2. **Loaded metal oxide modification**

   Loading metal ions is one of the more common modification methods. The metal oxides and hydroxyl oxides generated from the combination of oxygen-containing functional groups on the surface of biochar with metal ions can improve the adsorption performance of biochar for phosphate in water. The common used metal oxides include: silica, alumina, and titanium oxide. For example, Calcium and Magnesium modified biochar tobacco stem biochar adsorbs 13.9–49.7 times more PO_3^{4-} than unmodified biochar by loading MgO and $Mg(OH)_2$ etc. on the surface (Man and Yucheng 2018).

3. **Loaded organic modification**

   Loaded organic modification refers to loading a certain amount of organic substances or functionalized organic substances onto the surface of biochar to increase its number of functional groups and chemical affinity. The common used organic substances include: glucose, xylan, etc. For example, the two mineral materials were zeolite and vermiculite, straw biochar and palm biochar as research objects, and the organic compound β-cyclodextrin was loaded on the surface of these four materials to prepare new adsorption materials. The experimental results showed that the adsorption effect of materials bearing β-cyclodextrin was significantly improved; for the adsorption of Cd^{2+} in water, the maximum adsorption was straw biochar after loading β-cyclodextrin: 92.5 mg/g; the maximum adsorption of Pb^{2+} in water was zeolite after loading β-cyclodextrin: 159.6 mg/g (Shengyang 2017).

6.4 The Water Purification Mechanism of Modified Biochar

The adsorption mechanism of phosphorus by most modified biomass is physicochemical adsorption. Physical adsorption includes pore filling and electrostatic attraction, and chemical adsorption includes ion exchange, surface complexation, ligand exchange and chemical precipitation.

Physical adsorption is the formation of adsorption point on the surface of biochar to make the phosphorus element attached to achieve the effect of phosphorus removal. Chemical adsorption refers to the covalent bonding, coordination bonding and other chemical reactions, the phosphorus pollutants and adsorption materials to achieve the purpose of phosphorus removal.

In the specific experimental process, the water purification process of modified biomass materials is complex and diverse, not only sticking to one adsorption method, in most cases, several adsorption processes are involved in the process of phosphorus removal. For example, Fe^{3+} is reduced and immobilized in the form of nanoscale zero-valent iron on the strongly alkaline anionic resin, and a composite adsorbent is prepared to remove phosphate from the simulated wastewater and secondary effluent of the wastewater treatment plant. The composite adsorbent was prepared to remove phosphate from the secondary effluent of simulated wastewater and sewage treatment plants, because the nanoscale zero-valent iron would release Fe^{2+}, which would be oxidized to Fe^{3+}, and then reacted with phosphate to form $FePO_4$, which realized the selective adsorption of phosphate, and the process was due to electrostatic interaction, complexation and co-precipitation generated by phosphate adsorbed on the surface of the nanoscale zero-valent iron; moreover, the nanoscale zero-valent iron loaded on the resin effectively prevented the diffusion of humic acid into the resin through the physical pore limiting effect, reducing the potential binding between humic acid and cationic groups (Cong 2022).

Table 6.3 summarizes the main adsorption mechanisms of phosphate by biomass adsorbents modified by different methods.

6.5 Conclusion

Generally, the biochar obtained through modified preparation has good adsorption on nitrogen and phosphorus, and it has a brilliant potential of commercial application. However, there is still a space for improvement of modified biomass adsorbent in actual eutrophication water bodies. For example, the modification method can be more environmental-fridenly and more cost-effectively.

At the same time, we should pay attention to the relevant information of biomass aerogel and biomass sponge. These two kinds of biomass phosphorus removal materials are more environmentally friendly than modified biomass and are a major trend in the development of biomass materials in the future.

Table 6.3 Modification methods and adsorption mechanisms of biomass adsorbents

Raw materials	Modification method	Experimental condition	Theoretical optimum adsorption capacity/ (mg/g)	Adsorption mechanism	Bibliography
Reeds	Metal Load High Temperature Carbonization	Magnesium chloride solution 450 °C pyrolysis	317. 094	Ion exchange Physical adsorption Chemical adsorption	Cong (2022)
Soybean straw	Metal Load High Temperature Carbonization	Ferric chloride solution 700 °C pyrolysis	9.809	Ion exchange chemisorption	Mengjia (2018)
Walnut shell	Metal Load High Temperature Carbonization	$LaCl_3$ solution 400 °C pyrolysis	12.18	Chemisorption Intra-particle diffusion	Yuan et al. (2021)
Acorus calamus	Metal Load High Temperature Carbonization	$FeSO_4$ solution Pyrolysis at 673 °C	24.62	Surface adsorption, external membrane diffusion Intraparticle diffusion	Fengsheng et al. (2021)
Oyster shell	Metal loads	Ferric chloride solution Iron sulfate solution	9.81	Ligand exchange Electrostatic adsorption	Haoyan et al. (2021)
Digestate	Metal Load High Temperature Carbonization	$CaCl_2$, $MgCl_2$ solution 303 °C pyrolysis	76.92	Ion exchange chemisorption	Man et al. (2019)
Water hyacinth	Metal loading Co-impregnation-Pyrolysis	$CeCl_3$ solution	35	Ion exchange chemisorption	Guangliang et al. (2021)
Sludge	Metal loads	Chitosan and ferrous sulfate	49.32	Electrostatic attraction, pore filling, Surface chemical precipitation, Hydrogen bonding and ligand effects	Qianqian et al. (2021)

Acknowledgements Funding

This work was financially supported by National Key Research and Development Program of china (No.2023YFC3208702) and the Fundamental Research Funds for the Central Public Welfare Research Institutes (No. CKSF2021743/HL, No. CKSF2023311/HL), and the State-level Public Welfare Scientific Research Institutes Basic Scientific Research Business Project of China (No. CKSF2023337/SH).

References

Cong H (2022) Development of resin-based nano zero-valent iron phosphorus removal material and its application. Huazhong Agricultural University

Fengsheng W, Xiaoyi X, Hemin S et al (2021) Preparation of Fe^{2+}-modified calamus biochar and adsorption characteristics of phosphorus in water. J Environ Eng 15(11):3493–3503

Guangliang W, Wei Z, Shuaishuai L (2021) Adsorption characteristics of cerium-modified water hyacinth biochar on phosphate. Environ Sci 42(10):4815–4825

Haopu X, Bo T (2022) Comparative study on the effect of alkali-modified biochar in treating Cd^{2+}-containing wastewater. Environ Sci Manage 47(06):82–85

Haoyan M, Mingfan Y, Guozhi L et al (2021) Adsorption performance and mechanism of phosphorus in water by iron-loaded oyster shell powder. J Environ Eng 15(02):446–456

Jiali Z (2022) Adsorptive removal of cadmium and lead from water by sludge biochar loaded with magnesium and iron metal oxides. Guangdong University of Technology

Lin H, Peng Y, Zhiying L et al (2015) Study on the adsorption performance of magnetically modified diatomite on phosphorus in water. Environ Pollut and Prev 37(12), 41–44 + 50

Lulu W (2022) Degradation of sulfadiazine by oxygen-doped biochar-activated perovskite and reuse of elemental iron. University of Science and Technology of China

Man Y, Tingting L, Haihong L et al (2019) Adsorption characteristics of phosphorus in water by Ca/Mg-loaded modified digestate biochar. Environ Sci 40(03):1318–1327

Man Y, Yucheng C. (2018) Enhanced phosphate adsorption on Ca-Mg-loaded biochar derived from tobacco stems. Water Sci Technol 78(11):2427–2436

Mengjia T (2018) Research on the preparation modification of straw biochar and adsorption efficacy of nitrogen and phosphorus in water. Harbin Institute of Technology

Mengmeng Z (2019) Research on the preparation of modified biochar and its adsorption characteristics and mechanism of phosphorus. Xi'an University of Architecture and Technology

Oginni O, Yakaboylu AG, Singh K et al (2020) Phosphorus adsorption behaviors of MgO modified biochars derived from waste woody biomass resources. J Environ Chem Eng 8(2):103723

Qianqian S, Fangjun W, Yuantian Z et al (2021) Removal of phosphorus from water by iron-sulfur modified biochar. Environ Sci 42(05):2313–2323

Rivera-Utrilla J, Sánchez-Polo M, Gómez-Serrano V et al (2011) Activated carbon modifications to enhance its water treatment applications: an overview. J Hazard Mater 1:187

Shengyang Z (2017) Study of mineral and biochar materials loaded with β-cyclodextrin and adsorption of pollutants in water bodies. Yangzhou University

Shijie L, Hongsheng D, Jinhua S et al (2006) Discussion on water environment problems and water ecosystem restoration of lakes in China. China Water Resour 13:14–17

Sulin X, Congyuan G (2022) Progress of adsorption of phosphorus on metal-modified biochar. Appl Chem Eng 51(04):1088–1093 + 1100

Wenjie L, Panliang G, Shenghao S et al (2022) Research progress of phosphorus adsorption from wastewater by modified biochar. Shandong Chem Ind 51(06):137–138 + 145

Xiaoqiang C, Hulin H, Changkuan Z et al (2016) Capacity and mechanisms of ammonium and cadmium sorption on different wetland-plant derived biochar. Sci Total Environ 539:566–575

Xinfeng G, Hui P, Lianyi Z et al (2023) Lanthanum-modified biochar transported alone and co-transported with Cr(VI) in a sand column. J Agric Environ Sci 42(02):352–361

Xinran L, Guangguang G, Jing Z et al (2023) A review on the preparation of iron-modified biochar and its application to water remediation. Henan Sci 41(01)

Xuefei H, Zeqing W, Guanghua GE et al (2018) Current status of agricultural biomass adsorbents in water treatment. J Tarim Univ 30(04):76–82

Yanjun Z, Xintong L, Wenhui L et al (2021) One-step synthesis of garlic peel derived biochar by concentrated sulfuric acid: enhanced adsorption capacities for Enrofloxacin and interfacial interaction mechanisms. Chemosphere 290:133263

Yapen Y, Chaoyan Z, Junxian C et al (2021) Preparation of silkworm sand-based biochar by KOH activation and its adsorption characteristics on cadmium. J Environ Eng 15(11):3504–3514

Yuan C, Guangyao S, Chios CT et al (2006) Compositions and sorptive properties of crop residue-derived chars. Environ Sci Technol 38(17):4649–55

Yuan L, Kun X, Yiyang F et al (2021) Preparation and adsorption performance of lanthanum-modified walnut shell biochar for phosphate adsorption in water. Chem Ind Eng Prog 40(02):1121–1129

Yuhong Y, Lidong K, Qingfeng F et al (2022) Removal of phosphorus and antibiotics from water by magnesium-modified sludge-based biochar. China Environ Sci 42(09):4137–4144

Ziqiong S, Jianxin Z, Wei M et al (2022) Removal of phosphorus from water by acid/alkali modified cattail biochar and its mechanism. J Environ Sci 42(04):195–203

Chapter 7
Study on Groundwater Vulnerability Assessment of Rural Centralized Drinking Water Source

Shen Zhao, Ying Jiang, Ye Tian, and Xin Jiang

Abstract In the rural groundwater centralized water supply source vulnerability evaluation process, there are many difficulties in reflecting the process of fuzzy, uncertainty, and other issues. To solve these problems, this paper constructs a groundwater vulnerability evaluation model based on variable fuzzy theory. Taking the vulnerability evaluation of Huangqipu water supply sources as an example, the hierarchical analysis method was adopted to evaluate the water source from the perspectives of natural vulnerability and social vulnerability of the water source, respectively, and 13 first-level indexes were finally selected to construct the ecological vulnerability evaluation index system. The index system was evaluated qualitatively and quantitatively. The results show that the Huangqipu water source has high vulnerability and tends to be highly vulnerable. It is recommended to strengthen water quality monitoring and realize large-scale water supply as early as possible in areas with poor groundwater quality to ensure the safety of villagers' drinking water.

Keywords Variable fuzzy theory · Groundwater vulnerability assessment · Rural drinking water sources

S. Zhao
School of Water Conservancy and Environment, University of Jinan, Jinan 250022, Shandong, China

Y. Tian · X. Jiang (✉)
Water Resources Research Institute of Shandong Province, Jinan 250014, Shandong, China
e-mail: skyjiangxin@shandong.cn

Y. Jiang
Water Conservancy Comprehensive Service Center of Shandong Province, Jinan 250013, Shandong, China

Z. Wan (ed.), *Water Resources Management and Water Pollution Control*, Environmental Science and Engineering, https://doi.org/10.1007/978-3-031-53456-0_7

7.1 Introduction

Shandong is located on the eastern coast of China and in the lower reaches of the Yellow River. The agricultural population accounts for 36.95% of the total population. With the development of the rural economy, rural residents have higher and higher requirements for quality of life, and the rural centralized water supply project has become an essential public infrastructure for the comprehensive construction of a well-off society in rural areas. Because groundwater quality is relatively stable and the cost of exploitation is low, rural regions have formed a water supply structure dominated by groundwater sources and supplemented by surface water sources. Pollution of drinking water sources in rural areas is increasingly threatening the health of the rural regions (Singh et al. 2023). Less than 75% of rural centralized water supply projects in the province have water source protection measures, and less than 70% have emergency plans. The water quality of a few water supply projects has water quality problems such as high fluorine, bitter salt, and nitrate exceeding the standard (Huang et al. 2023). Scientifically assessing the vulnerability of rural drinking water sources is the basis for water source protection (Vogt and Sarkar 2015). There are various evaluation methods for groundwater vulnerability, mainly including groundwater pollution risk index method (Zhao et al. 2018), DRASTIC model (Shirazi et al. 2012), index-based vulnerability assessment models (Kumar et al. 2015), Integrated Aquifer Vulnerability Assessment (Jang et al. 2020), GOD model (Geng et al. 2023), pollutant migration processes and numerical simulation technology (Shrestha et al. 2017) and other methods. How to accurately evaluate the vulnerability of rural groundwater sources to take corresponding control and protection measures has important practical significance for promoting the standardized management of rural water supply projects and rural revitalization.

7.2 Study Area and Data

7.2.1 *Study Area*

The Huangqipu water supply source is near Huangqipu Town, Fangzi District, Shandong Province. It is located in the alluvial fan plain of Weiwen River, with an area of 482.50 km^2, about 33.0 km away from Weifang City. The terrain is relatively flat, slightly inclined north, and the slope is 2‰. The ground elevation is between 27.00 and 39.00 m above sea level. Affected by the monsoon circulation, the annual average precipitation in Huangqipu Street is 606.8 mm, and the rainfall is mainly concentrated in summer.

The Quaternary strata in the water source area are widely distributed. The groundwater is pore water. The lithology of the aquifer is mainly composed of fine sand, medium-coarse sand, and coarse gravel. The thickness of the aquifer ranges from 6.00 to 12.00 m. In recent years, the groundwater level of this water source area has risen slowly. The current water level is buried at a depth of 6.12 m, and the water production level is within the 12.00–20.00 m range. With the continuous increase of groundwater exploitation, artificial exploitation has become the primary discharge method of groundwater in the area. The dynamic changes of Quaternary pore water and groundwater in this area are mainly affected by precipitation, centralized exploitation of water sources, and seasonal agricultural exploitation.

The Huangqipu water supply source was suspended in 2014 due to the aging of the power distribution system and the severe leakage of supporting pipelines. To fully guarantee the water supply during the emergency, the local competent authority restored some wells in the Huangqipu water source area in 2016. In the first phase, 11 water intake wells near the Weihe River were restored, with a water supply scale of 20,000 m^3/d.

7.2.2 Data Source

The exploitable water resources in the study area, the annual average water resource exploitation, the buried depth of groundwater, the net recharge, and the aquifer medium are obtained from the Huangqipu Water Conservancy Station and local observation wells. The water quality monitoring data comes from 5 of the 11 groundwater source wells. According to the survey, domestic sewage in the area is directly discharged without treatment. The chemical fertilizer application intensity data are calculated using the Weifang City Statistical Yearbook and Shandong Province Statistical Yearbook. Water quality monitoring data are shown in Table 7.1.

Table 7.1 Wells water quality of the Huangqipu water sources

Sampling point	Well 1	Well 2	Well 3	Well 4	Well 5
Total hardness	831	330	345	661	786
Sulfate	173	105	180	180	151
Chloride	190	119	119	119	102
Manganese	0.025	0.025	0.025	0.025	0.025
Volatile phenol	0.001	0.001	0.001	0.001	0.001
Ammonia nitrogen	0.313	0.01	0.48	0.264	0.354
Total coliforms	225	150	4	0	0
Total number of colonies	21	6	5	36	39
Nitrate	50.5	4.2	74.2	71.2	62.1
Fluoride	0.115	0.478	0.457	0.157	0.58
Comprehensive pollution index	8.252	5.315	0.922	0.802	0.829

7.3 Indicators and Methods

7.3.1 Indicators

The vulnerability of rural centralized drinking water sources is a systematic concept, and its influencing factors include the natural environment, population, economy, and many other aspects. Compared with urban centralized drinking water sources, the monitoring conditions of rural centralized drinking water sources need to be more comprehensive. To objectively, accurately, and comprehensively reflect the vulnerability of different water sources, the following principles should be followed when determining the index system: (1) The principles of overall and representativeness. (2) Systematic and hierarchical principles. (3) Principles of scientificity and operability. (4) Quantitative and qualitative principles. (5) Comprehensive and categorical principles.

Based on the above index system, this study adopts the hierarchical classification method to evaluate the natural vulnerability and social vulnerability of water sources, and finally selects six secondary indicators and 13 primary indicators to construct the ecological vulnerability evaluation index system, and conducts a qualitative and quantitative evaluation of the index system, as shown in Table 7.2.

7.3.2 Methods

Variable Fuzzy Assessment Model. Variable Fuzzy Set is an improved fuzzy set used to solve the dynamics of fuzzy concepts and the inapplicability of the principle of maximum membership degree. It can also standardize qualitative and quantitative indicators to the same evaluation dimension. Since the vulnerability of water sources involves many factors, such as nature and society, its vulnerability evaluation system is typically complex, and the previous evaluation methods have problems such as difficulty in reflecting the fuzziness and uncertainty in the evaluation process (Jiang et al. 2018). Variable fuzzy set theory is developed based on relative membership, which can solve the above problems well.

The specific steps of the variable fuzzy evaluation model are as follows:

Let U be a pair of opposite fuzzy concepts $\tilde{A}$ and A^c universes, which u are arbitrary elements of U, $u \in U$. And $\mu_{\tilde{A}}(u)$ and $\mu_{A^c}(u)$ are the membership functions of this pair of opposite fuzzy concepts.

The $D_{\tilde{A}}(u)$ is called the relative difference of the u to the fuzzy concept $\tilde{A}$.

$$D_A(u) = \mu_A(u) - \mu_{A^c}(u) \tag{7.1}$$

$$\mu_A(u) + \mu_{A^c}(u) = 1 \tag{7.2}$$

Table 7.2 Indicators and criteria for vulnerability assessment of groundwater sources for centralized rural water supply

Indicators			High	High-Middle	Middle	Middle-Low	Low
Vulnerability of natural characteristics	Quantity of flow	Groundwater extraction rate (%)	> 120	110 ~ 120	100 ~ 110	85 ~ 100	< 85
	Engineering geology	Groundwater depth	9 ~ 10	7 ~ 8	5 ~ 6	3 ~ 4	1 ~ 2
		Net recharge	9 ~ 10	7 ~ 8	5 ~ 6	3 ~ 4	1 ~ 2
		Aquifer medium	9 ~ 10	7 ~ 8	5 ~ 6	3 ~ 4	1 ~ 2
	Water quality	Comprehensive pollution index	> 1	0.7 ~ 1	0.4 ~ 0.7	0.2 ~ 0.4	< 0.2
Vulnerability to social characteristics	Human activities	Sewage treatment rate (%)	< 60	60 ~ 70	60 ~ 71	60 ~ 72	> 90
		Fertilizer application intensity	> 1.2	1 ~ 1.2	0.8 ~ 1	0.6 ~ 0.8	< 0.6
		Water supply population (Ten thousand people)	> 10	5 ~ 10	1 ~ 5	0.1 ~ 1	< 0.1
	Engineering management	Engineering operations and maintenance	< 4	4 ~ 6	6 ~ 8	8 ~ 9	> 9
		Governing bodies and mechanisms	< 4	4 ~ 6	6 ~ 8	8 ~ 9	> 9
		User satisfaction	< 4	4 ~ 6	6 ~ 8	8 ~ 9	> 9
	Emergency response capacity	Monitoring and early warning capacity	< 4	4 ~ 6	6 ~ 8	8 ~ 9	> 9
		Emergency management mechanisms	< 4	4 ~ 6	6 ~ 8	8 ~ 9	> 9

Fig. 7.1 The position relationship between the attraction domain and the repulsion domain of X

Mapping:

$$\begin{cases} D_{\tilde{A}} : U \to [-1, 1] \\ u| \to D_{\tilde{A}}(u) \in [-1, 1] \end{cases} \tag{7.3}$$

This is called a variable fuzzy set.

$$\tilde{V} = \{(u, D)|u \in U, D_{\tilde{A}}(u) = \mu_{A^c}(u), D \in [-1, 1]\} \tag{7.4}$$

$$A_+ = \{u|u \in U, \mu_{\tilde{A}}(u) > \mu_{A^c}(u)\} \tag{7.5}$$

$$A_- = \{u|u \in U, \mu_{\tilde{A}}(u) < \mu_{A^c}(u)\} \tag{7.6}$$

$$A_0 = \{u|u \in U, \mu_{\tilde{A}}(u) = \mu_{A^c}(u)\} \tag{7.7}$$

In the formula, A_+, A_-, and A_0 are the main domain of $\tilde{A}$, the main domain, and the gradient boundary of $\tilde{A}^c$, respectively.

Suppose $X_0 = [a, b]$ is the domain of attraction of variable fuzzy set $\tilde{V}$, and $X = [c, d]$ contains $X_0 (X_0 \subset X)$ intervals.

Intervals are variable fuzzy sets $[b, d]$. Suppose $M \in [a, b]$, then $\mu_{\tilde{A}}(M) = 1$.

M can be set to a specific value or the median of the interval $[a, b]$ (Fig. 7.1).

$\forall x \in X$, If the $x \in [c, M]$, the relative membership function is as follows.

$$\begin{cases} D_{\tilde{A}}(u) = \left(\dfrac{x-a}{M-a}\right)^{\beta}; x \in \text{[a,M]} \\ D_{\tilde{A}}(u) = -((x-a)/(c-a))^{\beta}; x \in \text{[c,a]} \end{cases} \tag{7.8}$$

If the $x \in [M, d]$, the relative membership function is:

$$\begin{cases} D_{\tilde{A}}(u) = \left(\dfrac{x-b}{M-b}\right)^{\beta}; x \in \text{[M, b]} \\ D_{\tilde{A}}(u) = -\left(\dfrac{x-b}{c-b}\right)^{\beta}; x \in \text{[b,d]} \end{cases} \tag{7.9}$$

Let $A = B$, C be power exponents, and $D = 1$.

Assuming that there are m indicators, each indicator can be divided into h levels ($\beta = 1, 2, 3, \ldots c$), and the relative membership of each level is calculated by the following equation.

$$u'_{ih} = \left\{ \left[1 + \frac{\sum_{i=1}^{m} \left[w_i \left(1 - \mu_{\tilde{A}}(x_{ih})\right) \right]^p}{\sum_{i=1}^{m} \left[w_i \mu_{\tilde{A}}(x_{ih}) \right]} \right]^{\frac{\alpha}{p}} \right\}^{-1} \tag{7.10}$$

In the formula, u'_{ih} is the non-standardized relative membership degree; w_i is the index weight; α is the optimization criterion parameter, $\alpha = 1$ or 2 is the minimum one and two square criteria respectively; p is the distance parameter, and $p = 1 \vee p = 2$ represents the Hamming distance and Euler distance respectively.

The normalized relative membership function is shown in Eq. (7.11).

$$u_{ih} = \frac{u'_{ih}}{\sum_{h=1}^{c} u'_{ih}} \tag{7.11}$$

The eigenvalue of the level is calculated by Eq. (7.12).

$$H = \sum_{h=1}^{c} u_{ih} h, \quad H = [1, c] \tag{7.12}$$

The variable fuzzy evaluation model is calculated by Eq. (7.10) α and p. The judgment criterion of the variable fuzzy evaluation model is shown in Eq. (7.13).

$$\begin{cases} h = 1 \quad H \in [1, 1.5] \\ h = h \quad H \in [h - 0.5, h + 0.5] \ (h = 23 \ldots c - 1) \\ h = c \quad H \in [c - 0.5, c] \end{cases} \tag{7.13}$$

Determination of index weight. In this paper, the fuzzy weight of each influencing factor is determined by the combination of binary comparative fuzzy quantitative analysis method and three-level weight. First, the importance of the evaluation index is compared based on empirical knowledge, and the qualitative scale matrix E is given. Then, according to the consistency scale condition, matrix E is checked for consistency, and the sum of each row of matrix E is sorted from large to small to give a qualitative ranking of the importance of indicators. Finally, according to the relationship between the fuzzy tone operator and the relative membership degree, the relative membership degree of the index to the importance is determined, and the index weight vector is obtained after normalization.

7.4 Results

7.4.1 Weight

The fuzzy weight of each influencing factor is determined by the combination of binary comparative undefined quantitative analysis method and three-level weight. Next, take the three first-level indicators of engineering management as an example to solve the weights and construct the binary comparison matrix E. After the consistency test is performed on the matrix according to the consistency scaling condition, the matrix satisfies the consistency test condition.

$$\mathrm{E} = \begin{bmatrix} 0.5\ 0.5\ 0.3 \\ 0.5\ 0.5\ 0.3 \\ 0.7\ 0.7\ 0.5 \end{bmatrix}$$

The sum of each row of the matrix is 1.3, 1.3, and 1.9 in turn. The qualitative ranking of the importance of the indicators from large to small is: user satisfaction, project operation and maintenance, management organization, and mechanism. The non-normalized index weight vector is obtained according to the fuzzy mood operator and the relative membership relationship.

$$\mathrm{w}' = \begin{bmatrix} 0.429\ 0.429\ 1 \end{bmatrix}$$

Then the normalized indicator weight vector w is as follows.

$$\mathrm{w} = \begin{bmatrix} 0.231\ 0.231\ 0.538 \end{bmatrix}$$

7.4.2 Vulnerability Assessment Results

Substituting the relevant parameters into the formulas (7.10), (7.11), and (7.12), and obtaining the relative membership degree, standardized relative membership degree, and level feature value of the sample under the combination of the four parameters, and then according to the formula (7.13) for the sample. Grade evaluation, the vulnerability evaluation results of each sample under four different parameter combinations are shown in Table 7.4.

Fangzi District has a population of 72,900 living in the first-level water source protection zone and 79.15 km^2 of farmland and gardens. There are three food companies and beer companies, one centralized sewage treatment facility, and 25 pig and laying hen farms in the first-class protected area of the Huangqipu water supply source. Since the pore water in this area mainly receives infiltration recharge from atmospheric precipitation, upstream lateral runoff recharge, river seepage recharge, irrigation infiltration recharge, etc., all kinds of pollution sources in the study area will pose a threat to the protection of water sources. The above analysis shows that the groundwater vulnerability assessment results are more consistent with the actual situation (Tables 7.3 and 7.4).

Table 7.3 Vulnerability indicators and weights of Huangqipu water source

Indicators			Value	Weight			
Vulnerability of natural characteristics	Quantity of flow	Groundwater extraction rate (%)	98.25	1	0.301	0.6	0.181
	Engineering geology	Groundwater depth	7	0.402	0.367		0.089
		Net recharge	9	0.329			0.072
		Aquifer medium	8	0.268			0.059
	Water quality	Comprehensive pollution index	3.224	1	0.332		0.199
Social characteristics vulnerability	Human activities	Sewage treatment rate (%)	0	0.329	0.391	0.4	0.051
		Fertilizer application intensity	1.04	0.402			0.063
		Water supply population (Ten thousand people)	12.1	0.268			0.042
	Engineering management	Engineering operations and maintenance	7	0.231	0.32		0.030
		Governing bodies and mechanisms	8	0.231			0.030
		User satisfaction	8	0.538			0.069
	Emergency response capacity	Monitoring and early warning capacity	4	0.6	0.289		0.069
		Emergency management mechanisms	6	0.4			0.046

Table 7.4 Huangqipu water source vulnerability assessment results

	$\alpha = 1$ $p = 1$	$\alpha = 2$ $p = 1$	$\alpha = 1$ $p = 2$	$\alpha = 2$ $p = 2$	Mean	Vulnerability degree
Huangqipu water source	3.72	3.94	3.56	3.73	3.74	High-Middle degree tends to be the high degree

7.5 Conclusions and Suggestions

7.5.1 Conclusion

Based on the variable fuzzy evaluation model, a set of evaluation index system for groundwater source safety is established by using AHP, including two systems of natural characteristics and social characteristics. The evaluation system is simple and practical, and can fully demonstrate the interaction between human society and the environment from both qualitative and quantitative aspects. The key control indicators can guide the improvement of water source conditions. The evaluation results can support for comprehensive land use planning and groundwater resource protection planning in various regions.

Using the above index system and method, the vulnerability of the Huangqipu water supply source was evaluated. The results show that the average value of the characteristic value of the water source is 3.74, and the vulnerability of the water source is High-Middle and tends to be High. This example shows that the model method is simple to calculate, and can comprehensively and objectively reflect the vulnerability of shallow groundwater in actual areas through the changes of the model and its parameters α and p, and has high scientific and practical application value.

Although this study simulated the accurate results of the groundwater vulnerability assessment process based on variable fuzzy theory, there are still some limitations in data acquisition, evaluation parameter selection, single parameter scoring, and weight system construction.

7.5.2 Recommendations

According to the vulnerability assessment results of the Huangqipu water supply source, it is recommended to realize large-scale water supply in areas with poor groundwater quality as soon as possible to ensure rural drinking water safety. It should establish a dynamic water quality monitoring system, and constantly grasp the water quality status of water sources.

Acknowledgements This study was supported by the Open Research Fund of Henan Key Laboratory of Water Resources Conservation and Intensive Utilization in the Yellow River Basin (NO. HAKF202105) and the Optional Research Fund of Water Research Institute of Shandong Province (SDSKYZX202102).

References

Geng C, Lu D, Qian J, Xu C, Li D, Ou J, Ye Z (2023) A Review on process-based groundwater vulnerability assessment methods. Processes 11(6):1610

Huang S, Guo J, Xie Y, Bian R, Wang N, Qi H (2023) Distribution, sources, and potential health risks of fluoride, total iodine, and nitrate in rural drinking water sources of North and East China. Sci Total Environ 898:165561–165561

Jang WS, Engel B, Yeum CM (2020) Integrated environmental modeling for efficient aquifer vulnerability assessment using machine learning. Environ Model Softw 124:104602

Jiang X, Liu Y, Xu S, Qi W (2018) A gateway to successful river restorations: a pre-assessment framework on the river ecosystem in Northeast China. Sustainability 10(4):1029

Kumar P, Bansod BK, Debnath SK, Thakur PK, Ghanshyam C (2015) Index-based groundwater vulnerability mapping models using hydrogeological settings: a critical evaluation. Environ Impact Assess Rev 51:38–49

Shirazi SM, Imran HM, Akib S (2012) GIS-based DRASTIC method for groundwater vulnerability assessment: a review. J Risk Res 15(8):991–1011

Shrestha S, Kafle R, Pandey VP (2017) Evaluation of index-overlay methods for groundwater vulnerability and risk assessment in Kathmandu Valley, Nepal. Sci Total Environ 575:779–790

Singh K, Tewari G, Kumar S, Busa R, Chaturvedi A, Rathore S, Singh R, Gangwar A (2023) Understanding urban groundwater pollution in the Upper Gangetic Alluvial Plains of northern India with multiple industries and their impact on drinking water quality and associated health risks. Groundw Sustain Dev 21:100902

Vogt J, Sarkar R (2015) Drinking water vulnerability in rural coastal areas of Bangladesh during and after natural extreme events. Int J Disaster Risk Reduct 14:411–423

Zhao Y, Zhang J, Chen Z, Zhang W (2018) Groundwater contamination risk assessment based on intrinsic vulnerability, pollution source assessment, and groundwater function zoning. Hum Ecol Risk Assess Int J 25(7):1907–1923

Chapter 8
Revisited Coagulation-Flocculation-Nanofiltration for Dye Removal

Azreen Ibrahim, Nurul Syufiana Jumadil, Jonathan Fletcher Roger, and Abu Zahrim Yaser

Abstract Wastewater generated from industrial processes that contain dyes presents a significant environmental challenge, and traditional treatment methods often struggle to efficiently eliminate these harmful substances. Nanofiltration (NF) plays a crucial role in addressing dye wastewater treatment, but it encounters a major drawback in the form of fouling. To mitigate the fouling and improve the performance of NF for water reuse, coagulation-flocculation (CF) is suggested as a pretreatment step before NF. This review explores advancements after year 2011 in coagulation-flocculation-NF techniques for dye removal. Aluminum sulfate and ferric sulfate have been used as coagulants while polydiallyldimethyl ammonium chloride and anionic polyacrylamide have been used as flocculants. Application of natural coagulant-NF technique for dye containing wastewater treatment has potential to generate less sludge in comparison to alum, potentially reducing overall costs and promoting environmental sustainability. Moringa oleifera has potential as an alternative to the conventional coagulant. Synthesizing a natural coagulant through plant extraction for excellent coagulating agent needs to be explored in the future.

Keywords Natural coagulant · Modification · Polymer · Electrocoagulation

8.1 Introduction

A dye is typically defined as a substance with the ability to create color by forming physical or chemical bonds with the surface it is applied to. The color development is attributed to the presence of chromophores in the dye, which are connected to auxochromes (Dutta et al. 2021). The existence of residual dyes in surface water is visually unappealing and disrupts the aquatic ecosystem by diminishing sunlight penetration and depleting dissolved oxygen levels (Zahrim et al. 2011). Certain dyes

A. Ibrahim · N. S. Jumadil · J. F. Roger · A. Z. Yaser (✉)
Faculty of Engineering, Universiti Malaysia Sabah, 88400 Kota Kinabalu, Sabah, Malaysia
e-mail: zahrim@ums.edu.my

Z. Wan (ed.), *Water Resources Management and Water Pollution Control*,
Environmental Science and Engineering, https://doi.org/10.1007/978-3-031-53456-0_8

possess toxicity and mutagenicity, with the potential to release carcinogenic amines (Zahrim et al. 2011). Hence, it becomes imperative to safeguard the environment against the harmful dye and dye effluent discharges into water by subjecting them to various physical, chemical, biological treatments, or their combinations. Among these methods, NF is a noteworthy choice in dye wastewater treatment due to its superior permeate flux and low energy consumption compared to reverse osmosis (Zahrim et al. 2011). Nevertheless, the primary concerns with NF are irreversible fouling and scaling when applied directly. As proposed by Zahrim et al. (2011), the coagulation-flocculation method serves as an effective pretreatment step for treating the NF feed, with a specific focus on removing dye molecules among other contaminants. The efficiency of coagulation is contingent upon factors such as the molecular composition of the dye, its molecular weight, its ionic properties, and the accompanying chemicals used. Employing polymers as flocculant aids for the pretreatment of NF in the removal of highly concentrated dyes offers several benefits, including the reduction of sedimentation tank size, operation within a broader pH range, and improved removal efficiency when chemical auxiliaries are present.

In this paper, we revisit Zahrim et al. (2011) review article and explore contemporary developments and difficulties linked to coagulation-flocculation and NF for the removal of dyes from wastewater. The advancements in coagulation-flocculation and NF techniques for dye removal since year 2012 are summarized in Fig. 8.1. Excellent reviews on NF have been published by quite a number of authors on various aspects of NF membranes after the year 2011. Ahmad et al. (2022) discussed the NF applications in industrial wastewater treatment for water recycling, reuse, and product recovery including dye wastewater and mentioning fouling as one of the challenges that need to overcome. Jia et al. (2023) reported a key integration approach including coagulation-NF-based process using natural coagulant (tannin) for alleviating membrane fouling but this approach resulted in the increment of operating cost. Electrocoagulation is reported to reduce the cost of coagulant and Moneer (2023) has reviewed the roles of electrocoagulation as pre-treatment for membrane. Besides that, Naidua et al. (2021) stated that natural coagulant could have potential as an inexpensive, green, and effective coagulant as NF pre-treatment. Therefore, searching for various low-cost materials for natural coagulant is vital and will be emphasized in this review.

8.2 Coagulation as Pretreatment

Based on literature, the challenges of membrane fouling for wastewater processing can be effectively managed by the integrated CF-NF process. However, it should be noted that most of the studies are using chemical coagulants. The combination of CF and NF can complement each other's strengths and overcome their individual limitations. The treatments of dyes (acid, basic and reactive dyes) wastewater were studied by Liang et al. (2014). They applied individual coagulation/flocculation (CF) and nanofiltration (NF) processes as well as their combination (referred to as CF–NF).

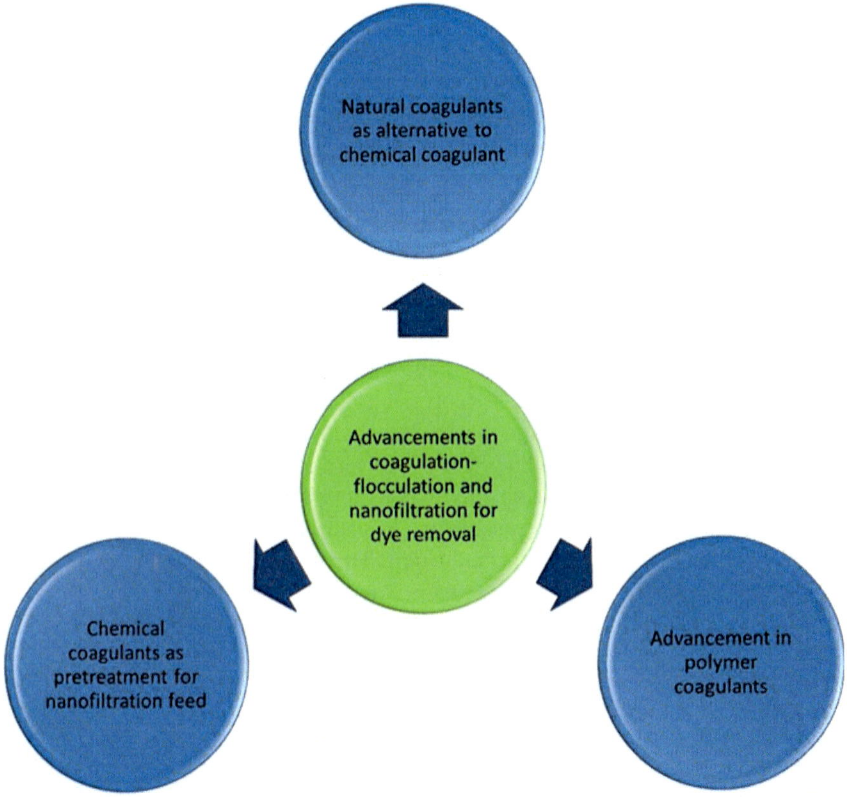

Fig. 8.1 Advancements in coagulation-flocculation and NF techniques for dye removal

For the treatment of highly concentrated multiple dyes wastewater (MDW), polyaluminum chloride (PAC) and polydiallyldimethyl ammonium chloride (PDDA) were found to be the most effective coagulant and flocculant, respectively. The CF process can achieve about 90% of dye removal at the optimal dosage of PAC/PDDA = 400/200 ppm, and the MDW with $pH > 3$ is favorable for the CF treatment. Similarly, Zahrim et al. (2010) also found that polydiallyldimethyl ammonium chloride was the best flocculant for Acid Black Dye 210 treatment. After coagulation-flocculation, Liang et al. (2014) using a positively charged NF hollow fiber membrane and the combination of CF-NF able to remove almost 100% dyes with a permeate flux of about 1.0 L m^{-2} h^{-1} under an operating pressure of 1 bar. The NF treatment can completely remove the strong color left in CF treated dye solutions, while the efficiency of coagulant/flocculant is improved by treating NF concentrated streams and subsequently results in much less sludge. In addition, membrane fouling is abated, and NF permeate flux is increased by applying the CF process as a pretreatment. Thus, the combination of CF–NF improves the overall performance for the dyes wastewater treatment (Liang et al. 2014).

In another study, Liang et al. (2015) reported that an effective CF formulation, i.e., ferric chloride/anionic polyacrylamide (IC/APAM) = 800/100 ppm, has been identified to dyes slurry treatment in a pilot plant. While the CF process can remove 91% of color and 81% of chemical oxygen demand (COD) from the slurry, the NF process is able to achieve 82 and 67% removal of color and COD from the colored wastewater, respectively. On the other hand, Çelebi et al. (2021) reported that coagulation using ferric chloride ($FeCl_3$) could enhance NF process. The method successfully removed calcium (Ca^{2+}), magnesium (Mg^{2+}), silica (SiO_2), and total organic content (TOC) up to 97, 83, 92, and 87% respectively.

Köse and Biroğul (2016) studied a coagulation-flocculation-membrane filtration process for treatment of real textile wastewaters. Aluminum sulfate and ferric sulfate were used as coagulants and several natural materials, namely limestone, magnesite, kaoline, pumice, and sedipür (polyelectrolyte), were used as flocculant aids. The experimental results showed that the treatment with aluminum sulfate (0.5 g/L) and ferric sulfate (0.18 g/L) at a pH of 6 was very effective. The color removal for aluminum sulfate and ferric sulfate reached 81 and 86%, respectively, and the COD was reduced by 70 and 47%, respectively. The natural flocculant aids behaved differently for color removal and COD reduction. The treatment with aluminum sulfate and limestone at pH 6 removed 89% of the color and reduced the COD by 80%. The treatment with ferric sulfate aided with natural materials at pH 6 did not significantly remove the color and reduced the COD value by 50%. Cellulose nitrate membranes can easily be cleaned using nitric acid.

8.3 Natural Coagulants as Alternative to Chemical Coagulant

Over the past decade, natural coagulants have been applied as an alternative to chemical coagulants for dye removal. Like an inorganic coagulant, a natural coagulant function by causing colloidal particles in wastewater to clump together, resulting in a reduced quantity of biodegradable sludge that is environmentally safe (Ibrahim et al. 2020). In a study conducted by Prabhakaran et al. (2020), a comparison was made among maize, green beans, nirmala seeds, and moringa seeds, in addition to the commonly used coagulant, alum. The findings revealed that all the natural coagulants effectively removed pollutants. Moringa seeds demonstrated remarkable performance by removing up to 90, 80 and 70% of total suspended solids (TSS), COD, and color respectively from the textile effluent, which surpassed alum. Furthermore, it generated less sludge in comparison to alum, potentially reducing overall costs and promoting environmental sustainability. Nevertheless, natural coagulants do come with challenges such as availability, cost, stability, and proper dosage. Similarly, study by El Gaayda et al. (2022) found that *Moringa oleifera* seed powder achieved 92.2% removal of Amido Black 10B dye from synthetic wastewater.

Other promising results of using natural coagulants derived from *Acacia nilotica* (babul pods) to treat textile wastewater was reported by Koul et al. (2022). The study showed 80 and 70% removal of turbidity and color respectively. However, future research on the need for pretreatment was addressed by the author, as efficiency depends on the wastewater being treated. The variability of natural coagulants also needs the study on the properties as every coagulant has different source of coagulant, method of extraction and the storage conditions which make it hard to optimize the use of coagulants.

Obiora-Okafo et al. (2022) investigated the potential of *Brachystegia eurycoma* and *Vigna subterranean* as natural organic polymers for the decolorization of Crystal Ponceau 6R (AR 44) in wastewater. The results showed that *Vigna subterranean* and *Brachystegia eurycoma* coagulants achieved optimal color removal rates of 88.8 and 73.3%, respectively. The kinetics of the coagulation-flocculation process were found to adhere to the pseudo-second-order model, with an R^2 value exceeding 0.997. This suggests that the mechanism controlling the rate is primarily governed by chemisorption.

The extraction of coagulants from Hibiscus sabdariffa seeds and the use of activated carbon through adsorption were explored by Hoong and Ismail (2018), for dye removal. The researchers identified that the pH of the solution had the most significant influence on dye removal efficiency, with optimal pH for effective dye removal being found to be 2. At this pH, the negatively charged dye molecules were more susceptible to aggregation with the coagulant and adsorption onto the adsorbent. The hybrid process's optimal conditions were determined as follows: pH 2, initial dye concentration of 385 ppm, coagulant dosage of 209 mg/L, and adsorbent dosage of 150 mg/L. Under these conditions, the dye removal efficiency reached an impressive 96.67%.

Murugaiyan et al. (2018) explored the potential of utilizing over 20 plant extracts and agricultural waste materials as natural coagulants for dye removal. Among these, neem, *Moringa oleifera*, and *Hibiscus sabdariffa* emerged as the most efficient plant extracts for dye removal, while rice husk ash, banana peels, and coffee grounds proved to be the most effective agricultural waste-based coagulants. Nevertheless, it is worth noting that there has been limited research into the combined use of these coagulants with nanofiltration.

8.4 Advancement in Polymer Coagulants

In a study by Wei et al. (2020), magnesium silicate polymer (MSP) an inorganic polymer was used for removal of reactive dyes such as Remazol Brilliant Blue R, Reactive Black 5, and Reactive Red 195. The coagulation behavior was observed to be influenced by its intrinsic pH, with a higher intrinsic pH resulting in improved dye removal. From the result, pH of 8.8 gave 70% removal of Reactive Black 5 but 90% removal of Remazol Brilliant Blue R was observed at pH of 10.2. The authors concluded that MSP exhibits great potential as a coagulant for effectively removing reactive dyes from wastewater and emphasized that MSP is environmentally friendly, with uncomplicated synthesis process. Nonetheless, it's important to note that MSP exhibits a certain degree of instability and can deteriorate over time, potentially impacting its efficacy in removing reactive dyes. Additionally, controlling the dosage of MSP proves to be quite challenging, with variations observed depending on the specific type of reactive dye used. Lastly, the cost associated with MSP is relatively high, which could present limitations in its widespread utilization.

In their comprehensive review, Ishak et al. (2020) explore the application of modified natural polymers for the removal of toxicant dyes, and they report notable improvements in their performance. The study focuses on three modified natural polymers which were guar gum, chitosan, and starch which were modified through processes such as grafting, cross-linking, and encapsulation to enhance their properties in removing toxic dye compounds, including azo, anthraquinone, and triphenylmethane dyes. The research emphasizes the potential of these modified natural polymers for efficiently removing toxic dyes from wastewater, highlighting their application-specific optimal parameters for achieving high removal efficiencies. The need for scalability, and further study on the biodegradable and toxicity properties were suggested for the modified natural polymers.

8.5 Conclusion

The fouling problem of nanofiltration membrane in dye removal from wastewater could be mitigated by applying coagulation-flocculation to pretreat the feed. The combination of CF and NF can complement each other's strengths and overcome their individual limitations during dye wastewater treatment. In recent years, a lot of advancements have taken place to improve the coagulants performance in removing dye efficiently i.e., application new synthetic polymer and limestones. Although researchers are moving towards electrocoagulation as pre-treatment, natural coagulants also have potential to be combined with NF process yet required further lab study.

References

Ahmad NNR, Ang WL, Teow YH, Mohammad AW, Hilal N (2022) Nanofiltration membrane processes for water recycling, reuse and product recovery within various industries: a review. J Water Process Eng 45:102478

Çelebi MD, Dilaver M, Kobya M (2021) A study of inline chemical coagulation/precipitation-ceramic microfiltration and nanofiltration for reverse osmosis concentrate minimization and reuse in the textile industry. Water Sci Technol 84(9):2457–2471

Dutta S, Gupta B, Srivastava SK, Gupta AK (2021) Recent advances on the removal of dyes from wastewater using various adsorbents: a critical review. Mater Adv 2:4497–4531

El Gaayda J, Titchou FE, Barra I, Karmal I, Afanga H, Zazou H, Yap P-S, Abidin Z, Hamdani M, Akbour RA (2022) Optimization of turbidity and dye removal from synthetic wastewater using response surface methodology: Effectiveness of Moringa oleifera seed powder as a green coagulant. J Environ Chem Eng 10:106988

Hoong HNJ, Ismail N (2018) Removal of dye in wastewater by adsorption-coagulation combined system with Hibiscus sabdariffa as the coagulant. MATEC Web of Conference 152:01008

Ibrahim A, Yaser AZ, Lamaming J (2020) Synthesising tannin-based coagulants for water and wastewater application: a review. J Environ Chem Eng 9:105007

Ishak SA, Murshed MF, Md Akil H, Ismail N, Md Rasib SZ, Al-Gheethi AAS (2020) The application of modified natural polymers in toxicant dye compounds wastewater: a review. Water 12:2032

Jia T-Z, Rong M-Y, Chen C-T, Yong WF, Lau SK, Zhou R-F, Chen M, Sun S-P (2023) Recent advances in nanofiltration-based hybrid processes. Desalination 565:116852

Köse TE, Biroğul NC (2016) Real textile wastewater reclamation using a combined coagulation/flocculation/membrane filtration system and the evaluation of several natural materials as flocculant aids. Gazi Univ J Sci 29(3):565–572

Koul B, Bhat N, Abubakar M, Mishra M, Arukha AP, Yadav D (2022) Application of natural coagulants in water treatment: a sustainable alternative to chemicals. Water 14:3751

Liang C-Z, Sun S-P, Li F-Y, Ong Y-K, Chung T-S (2014) Treatment of highly concentrated wastewater containing multiple synthetic dyes by a combined process of coagulation/flocculation and nanofiltration. J Membr Sci 469:306–315

Liang C-Z, Sun S-P, Zhao B-W, Chung T-S (2015) Integration of nanofiltration hollow fiber membranes with coagulation-flocculation to treat colored wastewater from a dyestuff manufacturer: a pilot-scale study. Ind Eng Chem Res 54(44):11159–11166

Moneer AA (2023) The potential of hybrid electrocoagulation-membrane separation processes for performance enhancement and membrane fouling mitigation: a review. Egypt J Aquat Res 49(3):269–282

Murugaiyan MGV, Ramesh M (2018) Application of locally sourced plants as natural coagulants for dye removal from wastewater: a review. J Mater Environ Sci 9(7):2058–2070

Naidua T, Qadir D, Nasir R, Mannan HA, Mukhtar H, Maqsood K, Ali A, Abdulrahman A (2021) Utilization of moringa oleifera and nanofiltration membrane to treat palm oil mill effluent (POME). Materialwiss Werkstofftech 52:346

Obiora-Okafo IA, Onukwuli OD, Igwegbe CA, Onu CE, Omotioma M (2022) Enhanced performance of natural polymer coagulants for dye removal from wastewater: coagulation kinetics, and mathematical modelling approach. Environ Processes 9:20

Prabhakaran G, Manikandan M, Boopathi M (2020) Treatment of textile effluents by using natural coagulants. Mater Today: Proc 33:3000–3004

Wei Y, Cheng X, Ding A, Xu J (2020) Magnesium silicate polymer as a coagulant for reactive dye removal from wastewater: considering the intrinsic pH in magnesium silicate polymer and coagulation behavior. ACS Omega 5:26094–26100

Zahrim AY, Tizaoui C, Hilal N (2010) Evaluation of several commercial synthetic polymers as flocculant aids for removal of highly concentrated C.I. Acid Black 210 dye. J Hazardous Mater 182(1–3):624–630

Zahrim AY, Tizaoui C, Hilal N (2011) Coagulation with polymers for nanofiltration pre-treatment of highly concentrated dyes: a review. Desalination 266:1–16

Chapter 9
Study on Eutrophication of Water Bodies Caused by Yangqu Reservoir Impoundment

Miaoxin Liu, Guoxin Xu, Quan Quan, Xingyu Liu, and Liting Tu

Abstract Yangqu Reservoir is a very important project in the construction of cascade hydropower stations in the upper reaches of the Yellow River. After the reservoir was built and put into operation, the rise of the water level in the upstream river caused an increase in the area of the water surface in the river, submerging the vegetation on both sides and reducing the self-purification capacity of the water body. The concentrations of COD, NH3-N and TN in the reservoir were simulated and predicted using the HD and AD models in MIKE21. The results showed that the concentrations of COD, NH3-N and TN in the reservoir had increased significantly. After the reservoir was built, on the one hand, due to the rise of the water level in the upstream river, the shedding and death of some submerged plants would cause local pollution concentration to exceed the standard, resulting in eutrophication of the water body. On the other hand, the hydraulic retention time increased, the degradation time of pollutants increased, and the change in environmental self-purification capacity was limited. To address the water quality issues of the reservoir, the spatial configuration (vertical and horizontal) and restoration design of vegetation communities were proposed to minimize the impact on vegetation communities caused by the construction of the reservoir, while also better playing multiple functions such as conserving soil and water, purifying water and maintaining biodiversity.

Keywords MIKE21 simulation · Yangqu Reservoir · Vegetation restoration · Water eutrophication

M. Liu · Q. Quan (✉) · X. Liu · L. Tu
State Key Laboratory of Eco-Hydraulics in Northwest Arid Region, Xi'an University of Technology, Xi'an 710048, Shaanxi, China
e-mail: qq@xaut.edu.cn

G. Xu
Hanjiang to Weihe River Valley Water Diversion Project Construction Co. Ltd, Xi'an 710199, Shaanxi, China

Z. Wan (ed.), *Water Resources Management and Water Pollution Control*, Environmental Science and Engineering, https://doi.org/10.1007/978-3-031-53456-0_9

9.1 Introduction

In recent years, the planning and construction of cascade hydropower projects in the upper reaches of the Yellow River have developed rapidly. The construction of the Yangqu Hydropower Station is of great significance for the development of water resources in the Yellow River Basin. In order to balance ecological protection, rational utilization of resources, and socio-economic development in the region, experts and scholars have conducted assessments of the ecological environment in this area and proposed reasonable plans for hydropower station construction. This plays a crucial role in promoting coordinated development between the regional ecology and economy. The Yangqu Hydropower Station is an integral part of the cascade hydropower stations in the upper reaches of the Yellow River. The construction of the Yangqu Hydropower Station has led to the rise of water levels in the reservoir area, resulting in the submergence of vegetation on both banks (Ji and Liu 2021). The submerged vegetation has lost its growth environment, leading to changes in the vegetation population and causing environmental issues such as eutrophication and pollution of the water body (Han and Yu 2016). The impact is irreversible. Therefore, the construction of the Yangqu Hydropower Station will have significant environmental effects on the upstream vegetation population, necessitating proper vegetation restoration measures.

9.2 Research Region Overview

9.2.1 Location

The Yangqu Hydropower Station (35°44′44″N, 100°16′46″E) is located at the junction of Xinghai County and Guinan County in Hainan Tibetan Autonomous Prefecture, Qinghai Province. The main purpose of the project is power generation and promoting local economic and social development. The area of the watershed above the dam site is 123,000 km^2, the average annual flow is 629 m^3/s, long-term average runoff of 1.9836×10^{10}m^3, reservoir designed normal water level 2,715 m, during operation, the normal water level, dead water level, and ecological limit water level are all 2,710 m, total storage capacity of the reservoir 1.639×10^9 m^3. The Yangqu Hydropower Station is located at the junction of Xinghai County and Guinan County in Hainan Prefecture, Qinghai Province, the distance from the upstream Banduo Hydropower Station is approximately 75 km. It is approximately 100 km from the downstream Longyangxia Hydropower Station and approximately 242 km from Xining City by road. It is the lowest level of the planned Ciha-Yangqu river section in the upstream of the main stream of the Yellow River Longyangxia Hydropower Station (Quan et al. 2018; Du 2017).

9.2.2 Inundation Range

With the help of Google Earth, the actual surface information was obtained and combined with the real situation of the region to determine the flooded area of the Yangqu Hydropower Station on the Yellow River (Source: "Environmental Impact Assessment Report of Yangqu Hydropower Station Project in 2020"). As shown in Fig. 9.1, the flooded area of the Yangqu Hydropower Station includes Xinghai County, Guinan County, and Tongde County, located in the low-lying areas.

9.3 Methods

The MIKE21HD and AD modules are used to simulate the hydraulic and water quality characteristics of Yangqu Reservoir. MIKE21 is a numerical modeling software developed by DHI, Denmark. It is widely recognized by researchers both domestically and internationally and has been proven to be highly accurate and efficient in various engineering projects (Yu et al. 2023).

In this study, we apply MIKE software to simulate river water levels, flow velocities, and water quality variations. The software can display the pre-reservoir water level and flow velocity characteristics as well as the changes in water quality before and after reservoir construction, enabling visualization of water levels, flow velocities, and water quality variations for further analysis.

9.3.1 Simulation Method of Hydraulic Models

The assumptions of the MIKE21 hydraulic module include the Boussinesq assumption and the hydrostatic pressure assumption. The governing equation for this module is the Navier–Stokes equation (Shi 2021).

Two-dimensional unsteady shallow water equations system:

$$\frac{\partial h}{\partial t} + \frac{\partial h\bar{u}}{\partial x} + \frac{\partial h\bar{v}}{\partial y} = hS \tag{9.1}$$

$$\frac{\partial h}{\partial t} + \frac{\partial h\bar{u}}{\partial x} + \frac{\partial h\bar{v}}{\partial y} = f\bar{v}\text{h-gh}\frac{\partial \eta}{\partial x} - \frac{h}{\rho_o}\frac{\partial Pa}{\partial x} - \frac{gh^2}{2\rho_o}\frac{\partial \rho}{\partial x} + \frac{\tau_{sx}}{\rho_o} - \frac{\tau_{\text{bx}}}{\rho_o} - \frac{1}{\rho_o}\left(\frac{\partial s_{xx}}{\partial x} + \frac{\partial s_{xy}}{\partial y}\right) + \frac{\partial (hT_{xx})}{\partial x} + \frac{\partial (hT_{xy})}{\partial y} + hu_s S \tag{9.2}$$

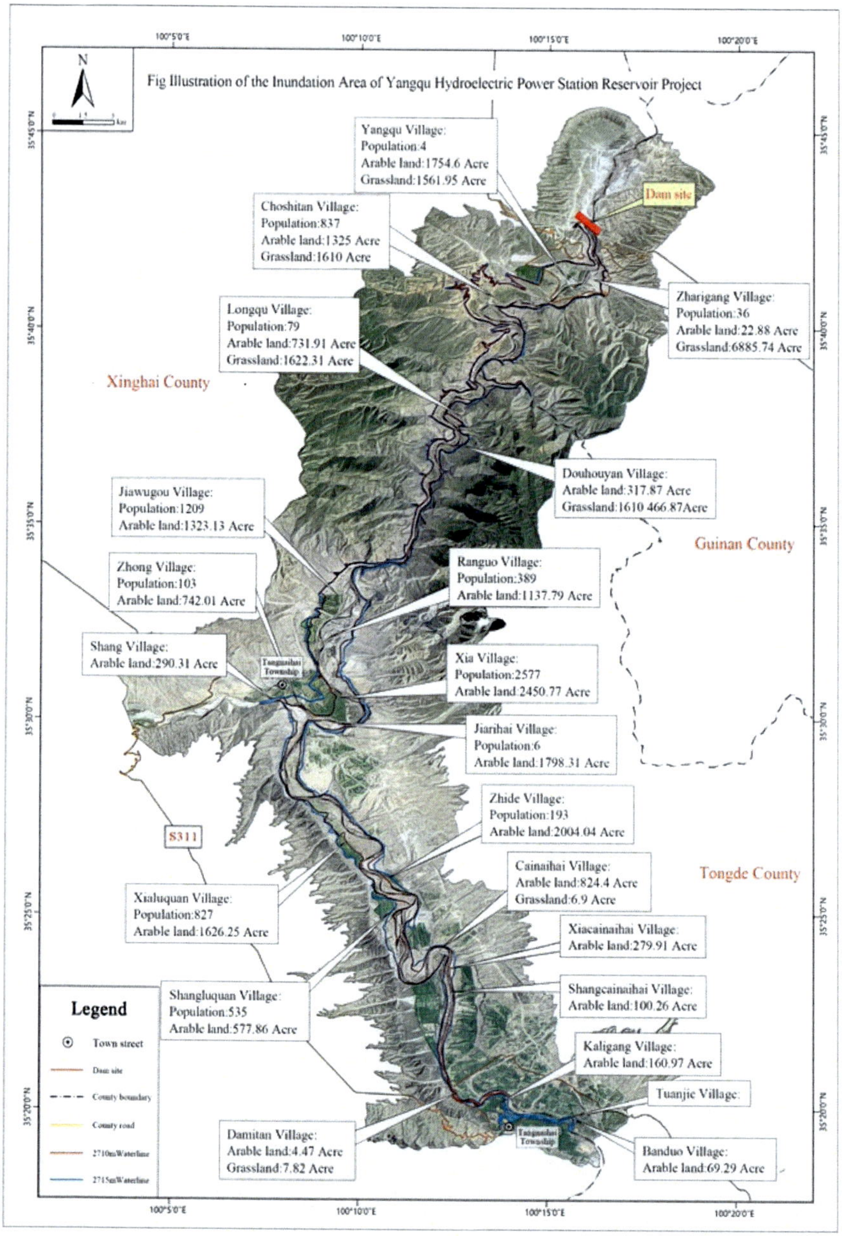

Fig. 9.1 Illustration of the inundation area of the Yangqu Hydropower Station Project

$$\frac{\partial h\bar{v}}{\partial t} + \frac{\partial h\bar{u}\bar{v}}{\partial x} + \frac{\partial h\bar{v}^2}{\partial y} = -f\bar{u}\text{h-gh}\frac{\partial \eta}{\partial y} - \frac{h}{\rho_o}\frac{\partial Pa}{\partial y} - \frac{gh^2}{2\rho_o}\frac{\partial \rho}{\partial y} + \frac{\tau_{sy}}{\rho_o} - \frac{\tau_{\text{by}}}{\rho_o}$$
$$-\frac{1}{\rho_o}(\frac{\partial s_{yx}}{\partial x} + \frac{\partial s_{yy}}{\partial y}) + \frac{\partial (hT_{xx})}{\partial x} + \frac{\partial (hT_{yy})}{\partial y} + hv_s S \quad (9.3)$$

where t, (x, y), η, and d separately represent time, coordinates in the Cartesian coordinate system, water level, and still-water depth; h is identical to $\eta + d$ and denotes the total water depth; u and v separately denote the velocity components in the x and y directions; f is the Coriolis force coefficient ($f = 2w \sin\varphi$); w, φ, g, and p separately represent the rotational angular velocity of the earth, local latitude, gravitational acceleration, and water density; S_{xx}, S_{xy}, and S_{yy} are components of radiation stress; S is the source item; is the water flow velocity of the source item. Symbols with a short line above represent mean values.

9.3.2 *Simulation Method of the Water Quality Model*

The water quality model is constructed by coupling the HD model (hydraulic model) and AD model (advection–diffusion model) (Quan et al. 2021).

1. Water quality control equation:

   Diffusion of pollutants in water is described by the advection–diffusion equation:

$$\frac{\partial C}{\partial t} + u\frac{\partial C}{\partial y} + v\frac{\partial C}{\partial y} = D_x\frac{\partial^2 C}{\partial x^2} + D_y\frac{\partial^2 C}{\partial y^2} \quad (9.4)$$

where C is concentration (mg/l); D_x and D_y separately denote the diffusion coefficients in x and y directions.

2. Water quality degradation equation

   Degradation of pollutants in water is modeled by a first-order reaction equation:

$$\frac{\partial C}{\partial t} = -KC \quad (9.5)$$

where t, C, and K denote time (s), concentration (mg/l), and attenuation coefficient (s^{-1}), respectively.

9.4 Results and Discussions

9.4.1 *Hydraulic Simulation Results of Yangqu Reservoir*

Prior to the construction of the Yangqu Reservoir, the hydrodynamic characteristics of water level and flow velocity in the upstream river channel were analyzed. The water level and flow velocity distribution maps for the upstream river channel are available for the months of July, October, January, and April (see Figs. 9.2 and 9.3).

According to Fig. 9.2, it can be observed from the water level distribution map of Yangqu Reservoir before its construction that under the natural river channel conditions, the water level remained relatively stable between 2,695 and 2,635 m. During the flood season from mid-September to early October, the water level at the downstream of the reservoir was slightly elevated due to inflows from upstream, but the magnitude was not significant. During the non-flood season with low flow rates, the water surface remained stable.

According to Fig. 9.3, it can be observed from the velocity distribution map of the natural river channel in Yangqu that the water flow velocity is relatively fast. During the non-dry season, the velocity varies within the range of 3.0 m/s–0.5 m/s, depending on the width of the cross-section. During the dry season, the flow velocity significantly decreases due to reduced inflow, but the overall velocity in the reservoir area remains above 0.5 m/s.

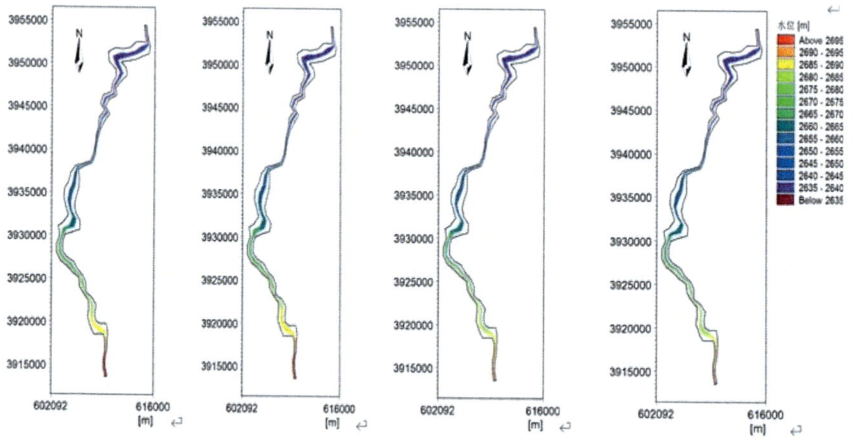

Fig. 9.2 Water level distribution maps of Yangqu Reservoir before its construction (under natural river channel conditions) are shown from left to right for the months of July, October, January, and April

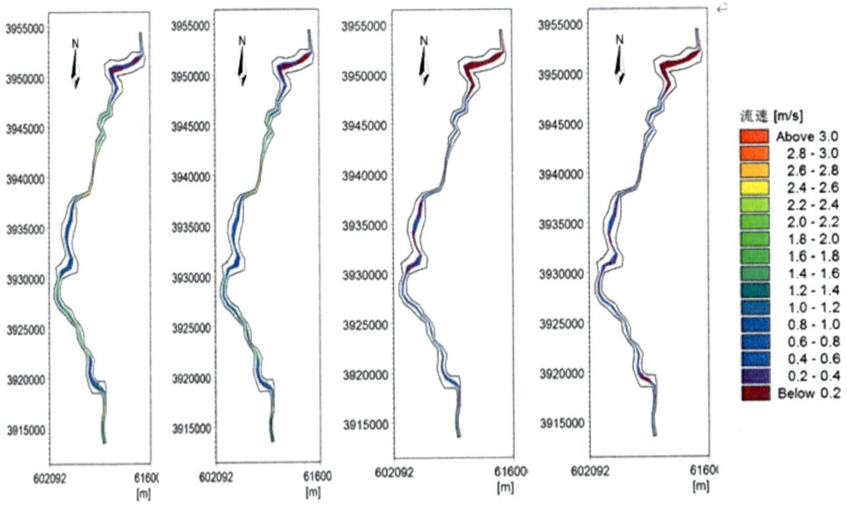

Fig. 9.3 Velocity distribution maps of Yangqu Reservoir before its construction (from left to right: July, October, January, April)

9.4.2 *Simulation Results of Water Quality in Yangqu Reservoir*

Based on the hydrodynamic model of Yangqu Reservoir, a water quality prediction model was constructed. The simulated water quality indicators include COD, NH3-N, and TN, which represent the variations in water quality for the natural river channel of Yangqu in February and July. The simulation results are shown in Figs. 9.4–9.6. After the construction of the reservoir, the surface area of the water body in Yangqu increased, the flow velocity decreased, and the degradation rate of pollutants slowed down. On the other hand, the hydraulic retention time increased, leading to longer degradation time for pollutants. However, the change in self-purification capacity of the environment was limited. Since the mainstream flow is larger than the inflow from tributaries, the water quality in the river section is predominantly influenced by the water quality of the upstream inflow.

From Figs. 9.7 to 9.9, it can be observed that the concentrations of various pollutants changed over time before and after the construction of the reservoir. Due to abundant rainfall in the summer, soil erosion increased, resulting in elevated levels of various pollutant indicators during this season. At the dam site, the concentration of COD_{Mn} increased from 1.06 mg/l to 2.92 mg/l. Simultaneously, ammonia nitrogen increased from 0.05 mg/l to 0.09 mg/l, and TN increased from 0.71 mg/l to 1.13 mg/l. Due to the increased hydraulic retention time, the peak arrival time at the Tangnaihai monitoring station and reservoir dam site was delayed. With the degradation of pollutants over a longer period of time, there was a slight improvement in water quality indicators compared to before the reservoir construction.

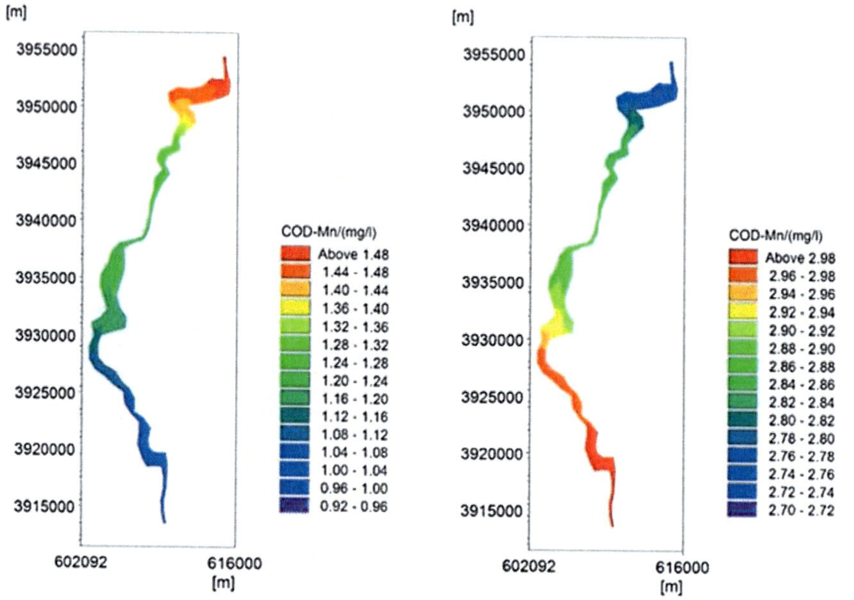

Fig. 9.4 COD_{Mn} water quality in Yangqu Reservoir in February (left) and July (right) after its construction

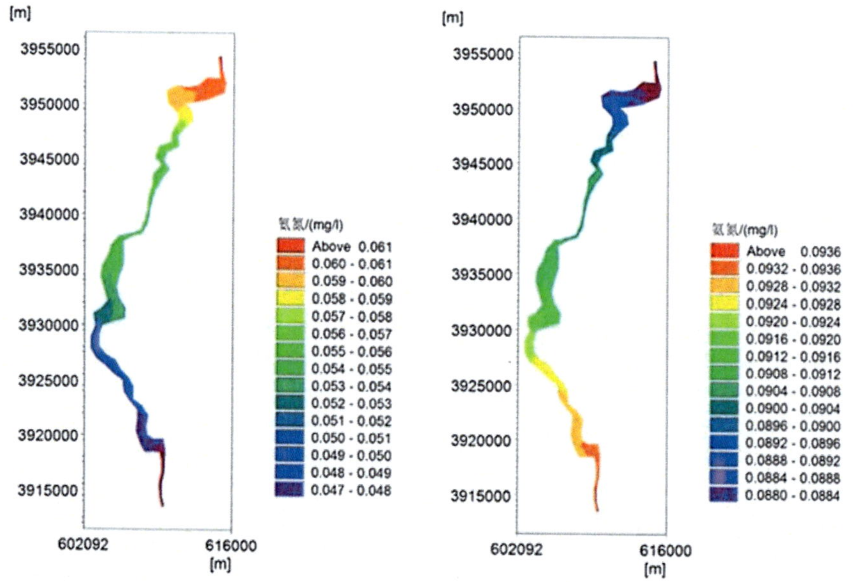

Fig. 9.5 Ammonia nitrogen (NH3-N) water quality in Yangqu Reservoir in February (left) and July (right) after its construction

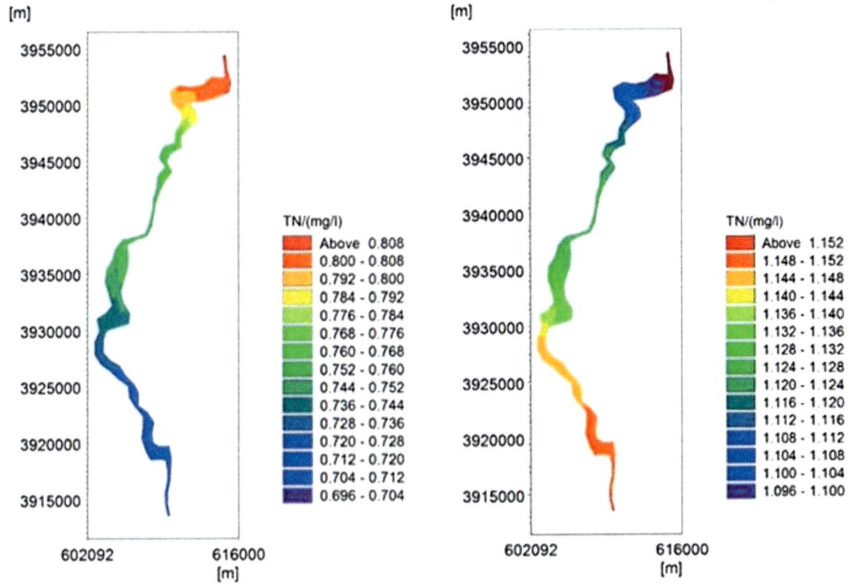

Fig. 9.6 TN (total nitrogen) water quality in Yangqu Reservoir in February (left) and July (right) after its construction

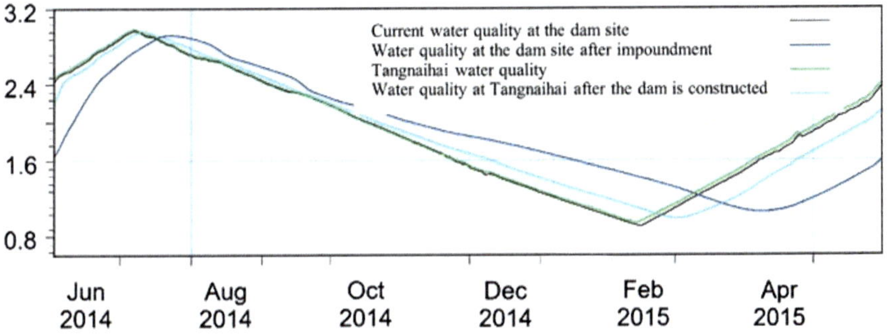

Fig. 9.7 Change in COD (unit: mg/l) before and after the construction of the reservoir

Based on the hydrodynamic model, this chapter constructed a water quality model for the current status of Yangqu Reservoir after its construction. The model was run and calculated using the Transport module of the MIKE software series. The following conclusions were drawn:

1. Main channel has a large flow rate, while the inflow from tributaries is relatively small. Therefore, the water quality in the river section is controlled by the quality of upstream incoming water.
2. Different types of pollutants fluctuate within a certain range depending on the season. At the dam site of Yangqu Reservoir, COD_{Mn} varied within the range

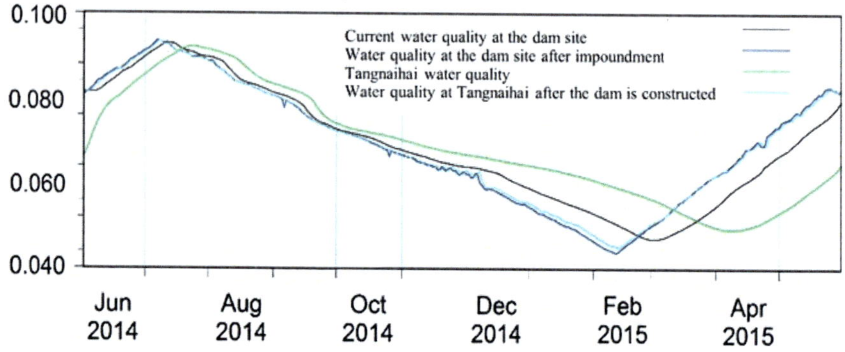

Fig. 9.8 Change in ammonia nitrogen (unit: mg/l) before and after the construction of the reservoir

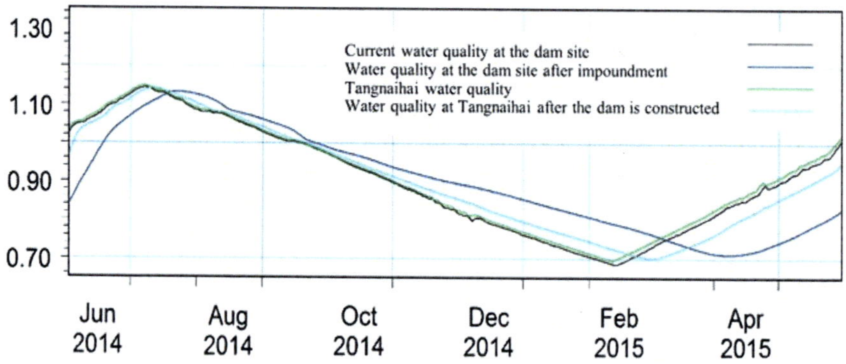

Fig. 9.9 Change in TN (total nitrogen) (unit: mg/l) before and after the construction of the reservoir

of 1.0 mg/l to 3.0 mg/l, ammonia nitrogen varied within the range of 0.05 mg/l to 0.10 mg/l, and TN varied within the range of 0.7 mg/l to 1.2 mg/l. During the summer season, with abundant rainfall, the various pollutant indicators enter the river with runoff and generally show a trend of being higher in summer and lower in winter.

3. Compared to before the construction of the reservoir, the water surface area in the Yangqu water area has increased, the flow velocity has decreased, and the degradation rate of pollutants has slowed down. On the other hand, the hydraulic retention time has increased, leading to longer degradation times for pollutants and limited changes in the self-purification capacity of the environment. Due to the increased hydraulic retention time, the peak arrival time at the Tangnaihai monitoring station and the reservoir dam site has been delayed. As pollutants degrade over a longer period, various water quality indicators have slightly improved compared to pre-construction conditions.

9.5 Constructing a Vegetation Restoration Design Plan

As concluded in Chap. 4, the construction of the Yangqu hydropower station resulted in the inundation of vegetation in the riparian zone due to the elevation of the water level. The loss of growth environment caused damage to the existing vegetation ecosystem, and the degree of impact is irreversible. After the construction of the reservoir, the pollution concentrations of three water quality indicators (COD_{Mn}, ammonia nitrogen, and TN) in the river channel were significantly higher than those in the pre-construction area. Considering the possibility of eutrophication occurring in some areas of the river, the survival of vegetation may be affected, exacerbating water quality degradation. Therefore, we can consider planting a certain number of vegetation communities on both sides of the riverbank that are more affected by water flow, to alleviate the impact on upstream river channel water level, flow rate, and water quality, and better achieve the goals of water conservation, soil retention, reducing water erosion, and improving water purification capacity. Under the premise of ensuring the stability of the vegetation community, we can better maintain the ecological stability of the study area. This chapter summarizes the spatial configuration method of vegetation communities and restoration design plan based on the vegetation problems in the Yangqu reservoir area and experimental results from previous studies.

9.5.1 Existing Vegetation Communities in the Yangqu Reservoir Area Face the Following Problems

October 12–14, 2023, Remote sensing field sampling within the inundation area of the Yangqu Hydropower Station, On-site survey was conducted at 18 sampling points as shown in Table 9.1.

Based on the 2020 Land Survey Bulletin of Qinghai Province, it is known that the forest land in the study area includes deciduous forests, shrub forests, and other types of forests. Deciduous forests account for 14.65%, shrub forests account for 80.24%, and other types of forests account for 5.11%. According to the results of on-site research and exploration, the local vegetation has the following issues, specifically: (1) Vegetation community has a single species and lacks prominent dominant species; (2) The shrub vegetation is abundant, but it consists of a single species. Tree species, grasslands, and aquatic plants are relatively scarce in the upstream river channels. (3) Due to the construction of the Yangqu Reservoir, water levels on both sides will rise, posing a serious threat to the survival of vegetation along the banks. (4) In recent years, desertification has become a growing concern due to natural environmental changes and excessive grazing. Currently, this problem is compounded by the potential impact of the Yangqu hydropower station on water levels, flow rates, and water quality, which could further disrupt plant population structures, damage local biodiversity, and undermine ecological stability.

Table 9.1 Survey point information

Number	Long	Lat	Elev (m)	Point information
1	100°16.7882″E	35°44.4556″N	2,435	To the northeast of the Yangqu Hydropower Station, there are solar panels to the north, a river to the east, a breeding station to the west, and a bald mountain directly facing the west, with sparse grass on the grassland and thin grass cover on the ground
2	100°17.4407″E	35°44.1531″N	2,515	To the east of the hydropower station under construction on the opposite bank: terraced fields covered with green nets on the stones, with only a few grass patches around, and the rest covered with small stones
3	100°17.6506″E	35°44.1068″N	2,617	To the northeast of the hydropower station under construction on the opposite bank: mainly bald mountains, with sparse grassland
4	100°15.9261″E	35°41.9493″N	2,894	A mountain composed of rocks, without forest, with a few shrubs, and sparse grassland distributed on the bald mountain
5	100°15.0835″E	35°42.0181″N	2,844	Woodland with shrub distribution
6	100°14.4908″E	35°41.6482″N	2,787	Arable land, primarily planted with corn
7	100°14.9424″E	35°41.5297″N	2,748	Next to the arable land, there is woodland primarily consisting of white poplar and elm trees
8	100°15.1257″E	35°40.7743″N	2,673	Gama Yangqu Bridge, surrounded by reeds, with relatively dense grassland growth
9	100°14.3294″E	35°40.9613″N	2,701	The land was originally arable, but now there are no crops growing, forming grassland
10	100°13.3417″E	35°40.7222″N	2,688	The area of arable land that has now been converted into grassland is located next to the temple, with many village houses built of earth nearby
11	100°10.2814″E	35°43.4528″N	3,001	Arable land, primarily cultivated with wheat and barley
12	100°08.9600″E	35°43.6120″N	2,792	Replanting area for willow trees: the location on the willow-lined bank
13	100°08.6302″E	35°32.5251″N	2,737	dwarf poplar
14	100°08.9802″E	35°32.6247″N	2,727	Shangjiawu Village: The areas to the northeast and northwest are primarily grassland, while the southern area is woodland
15	100°17.4709″E	35°47.5065″N	–	Jiarihai Village: The main crop type is corn
16	100°12.3023″E	35°25.4911″N	2,853	Benlong Village: The main crop type is corn

(continued)

Table 9.1 (continued)

Number	Long	Lat	Elev (m)	Point information
17	100°15.8792″E	35°19.9364″N	–	The construction of the Bando Hydroelectric Power Station affects the farmland on the opposite bank: The main crops grown are barley and corn
18	100°08.5467″E	35°30.9170″N	2,735 m	Tangnaihai Township: The main crops grown on the farmland include corn, wheat, and potatoes

During the field survey and research, various photos of different types of landforms were taken, including representative photos of cultivated land, forests, and grasslands. As shown in the following Figs. 9.10–9.13

Fig. 9.10 Farmland near the hydroelectric power station

Fig. 9.11 Grassland near the hydroelectric power station

9.5.2 Principles of Vegetation Arrangement

In order to address the issues present in the vegetation community within the study area, it is important to fully consider the following principles of vegetation arrangement when developing a vegetation restoration plan (LY/T 3316–2022; Mao et al. 2014; Mu et al. 2008).

1. Possess a moderate level of tolerance to both flooding and drought;
2. Have a well-developed root system and strong soil retention capacity;
3. Exhibit high survival rates and strong germination capabilities;
4. Demonstrate strong water purification abilities;
5. Native plants should be the primary species with supplementary exotic plants;
6. Prioritize protection, prioritize natural restoration, and combine artificial and natural restoration methods while adhering to the inherent laws of the ecosystem in order to enhance the self-repair and stability of the ecosystem.

Fig. 9.12 Forest near the hydroelectric power station

9.5.3 Vegetation Community Spatial Arrangement and Restoration Design

The main vegetation ecological types in the source area of the Yellow River include alpine shrubs, alpine meadows, alpine grasslands, marshes, riparian wetland plants, and aquatic vegetation. Among them, alpine meadows can be further classified into typical alpine meadows, steppe-like meadows, and marsh-like meadows. Experimental studies conducted by Xie et al. (2021), Zhang et al. (2019), Zhao et al. (2023), and Wang (2005) have found that emergent aquatic plants such as reeds, cattails, floating grass, calamus, canna, and umbrella grass have high purification capabilities for water quality. When configuring vegetation communities, it is important to consider both the vertical and horizontal arrangement. The vertical arrangement is mainly related to sunlight, with tall trees at the highest position, shrubs and herbaceous plants in the middle, and emergent aquatic plants at the lowest level. The horizontal arrangement is influenced by factors such as topography and human activities. According to the characteristics of the river channel, aquatic plants, herbaceous plants, shrubs, and trees are arranged sequentially from the riverbank to the land. Proper configuration of various vegetation types helps improve the ecological stability of vegetation communities.

Fig. 9.13 Shrubbery near the hydroelectric power station

Considering the water level fluctuation in the river channel during the filling of the Yangqu Reservoir ranging from 2,710 to 2,715 m, it can be divided into two sections. The specific division and restoration design are as follows:

Part 1: (2,710–2,713 m): Within this range, the water level remains high for a long time, and faster water flow is required. Therefore, emergent plants are planted, considering the harsh geographical conditions of the study area, which require tolerance to cold climates and flooding. The plant species include reeds, cattails, and calamus, but to increase plant diversity, flood-tolerant herbaceous plants such as water lilies and sedges are also added in moderate amounts.

Part 2: (2,713–2,715 m): Within this range, the water level does not submerge the vegetation. Therefore, shrubs and trees can be planted, but due to the difficulty of tree survival in Qinghai, local forest vegetation is chosen. The plant species include white willow, elm, Populus davidiana, Chinese tallow tree, and shrub roses.

References

Du Y (2017) Study on reservoir regulation of cascade hydropower stations in diversion projects. Northwest Hydropower 2017(3):62–64+76

Han M, Yu HZ (2016) Wetland dynamic and ecological compensation of the Yellow River Delta based on RS. Energy Proc 104

Ji YB, Liu ZQ (2021) The influence of three Gorges Reservoir water storage on the soil of the reservoir area. Zhejiang Water Res Sci Technol 49(5):1–11+15

LY/T 3316–2022, Technical specifications for vegetation restoration in small watersheds on the Loess Plateau

Mao J, Guan C, Huang WH et al (2014) A brief discussion on the ecological restoration techniques for lakes and rivers. Res Conserv Environ Protect 1:155

Mu J, Li ZB, Li P et al (2008) A preliminary study on the ecological restoration techniques of the reservoir area's disappearing zone in the Jinsha River dry and hot valley hydropower station. Bull Soil Water Conserv 28(6):172–176

Quan Q, Gao SZ, Shang YW, Wang BX (2021) Assessment of the sustainability of gymnocypris eckloni habitat under river damming in the source region of the Yellow River. Sci Total Environ 778:146312

Quan Q, Wang Y, Tian KD et al (2018) The impact and protection of indigenous fish by the Yangqu Hydropower Station in the upper reaches of the Yellow River. Environ Impact Assess 40(6):63–66

Shi X (2021) Study on the water quality variation characteristics and pollutant flux of the Si River based on the MIKE21 hydrodynamic-water quality coupled model. Shandong University

Yu Y, Cui XY, Ma ZZ (2023) Evaluation of flood control capacity in the urban section of Hunchun River using a one-dimensional model and MIKE21 model. Northeast China Water Res Hydropower 41(9)

Wang Y (2005) Study on the purification capacity of aquatic plants for urban domestic wastewater. Sichuan Agricultural University

Xie YB, Quan Q, Zou H et al (2021) Meta-analysis of the purification capacity of emergent aquatic plants in constructed wetlands. China Soil Water Conserv 8:50–53

Zhang QN, Chen YH, Yang HR et al (2019) Study on the purification capacity of 29 aquatic plants for rural domestic wastewater. J Agric Res Environ 36(3):392–402

Zhao XH, Fu Y, Guo Y et al (2023) Comparison of net water plant selection and purification capacity in the upper reach of Guixi River. Environ Pollut Cont 45(8):1101–1107

Chapter 10
Assessment of the Effect of Illumination on the Survival Rate of Eichhornia for Mine Water Treatment in Winter Time

Vladimir Dmitrienko, Irina Kokunko, Nadezhda Dmitrienko, and Nikolai Klavdiev

Abstract After the liquidation of coal enterprises in eastern Donbass, a large volume of mineralised mine water had to be pumped to the surface. The importance of qualitative treatment of pumped water out of old mine workings is given. The results of analyzing the efficiency of mine water treatment in the town of Shakhty, Rostov region, are given. Based on the analysis of the chemical composition of discharged water, the increased concentration of salts was determined. Some negative effects of discharging of treated groundwater into small rivers of the region were stated. A chance of water post-treatment purification by using Eichhornia is considered. The results of measuring natural illumination during different periods of the year are presented. The minimum permissible duration and illumination of plants in winter were determined experimentally in laboratory conditions. The results of evaluation of Eichhornia survival rate of during unfavorable winter periods are presented. The conditions for effective vegetation of a tropical plant to reduce the period of adaptation in the natural environment are noted.

Keywords Mine water · Pollution · Chemical composition · Post-treatment of waste water · Eichhornia · Vegetation

10.1 Introduction

Water is an integral part of life on Earth. At the same time, the share of fresh water contained in surface reservoirs is only 3.5% of the required total volume, but the available volumes of water are even smaller. Natural disasters and human activities have adversely affected the provision of high-quality fresh water to the population of

V. Dmitrienko (✉) · I. Kokunko · N. Dmitrienko · N. Klavdiev
Institute of Service and Business (Branch) of DSTU in Shakhty, 346500 Shakhty, Rostov Region, Russia
e-mail: vadmitrienko@rambler.ru

Z. Wan (ed.), *Water Resources Management and Water Pollution Control*, Environmental Science and Engineering, https://doi.org/10.1007/978-3-031-53456-0_10

many countries (Savenko et al. 2012). Therefore, the task of preserving this important natural resource is one of the most important goals of humanity (Lashkov 2011).

The next important component of water supply is groundwater that occupies the second position of the world ocean. However, this resource is limited and many countries are already facing this problem (Smirnova and Pozdnyakova 2013).

The mining industry, with its technological processes including pumping of huge volumes of groundwater to the earth surface, has a significant impact on quality and availability of groundwater (Agapov et al. 2007).

10.2 Environmental Problems of Eastern Donbass

Dozens of coalmines operated on the territory of eastern Donbass in the last century. Due to their closure and subsequent flooding of mine workings, the groundwater level reached critical values and could cause flooding of some populated areas (Bespalova et al. 2006). As a result, the main purpose was to pump water from mine workings. On the territory of three former mines in town of Shakhty, the process of pumping of underground water to the surface was carried out with the capacity of up to 1,500 m^3/hour. This is almost twice higher than all domestic wastewater of the town.

Paying attention to the need to lower the groundwater level, it was necessary to clean the polluted mine waters. As the discharge of untreated, mine waters would lead to extreme pollution of small rivers in the region. The organization of treatment facilities has significantly reduced the content of solid particles and dissolved iron, but as to salt solutions, they have been discharged into surface water reservoirs for years.

The mining of coal seams was accompanied by the destruction of rocks and, accordingly, the formation of reservoirs with intensive filtration of groundwater. As a result, water is getting to be saturated with a great quantity of dissolved substances (Shchadov et al. 2003; Demkin et al. 2009). Thus, chemical composition of the treated mine waters shows a significant content of magnesium, calcium, iron and other substances that can lead to anthropogenic pollution when discharging them into small rivers (Kaplunov et al. 2011).

In accordance with the results of the study of the chemical composition of groundwater in the town of Shakhty, Rostov region, it turned out that the concentration of hydrogen ions in pH is from 5.2 to 6.6, which makes it possible to characterize them as acidic on the acidity scale. Acid groundwater composition usually formed because of the oxidation of sulfur sulfides (pyrite) contained in their composition under the influence of oxygen dissolved in groundwater. As a rule, groundwater from mine workings previously owned by the Rostov-Coal Production Association is characterized as sodium chloride and has increased mineralization, in some cases exceeding 20 g/l (Kulikova et al. 2020). The temperature of the mine drainage varies from 6 °C to 25 °C all year round and depends on the depth of the workings.

The existing technologies of groundwater treatment at former mines are designed to remove solid mineral particles and free iron, but they do not allow for quality purification of substances dissolved in mine water. The application of other methods and means for mine water treatment in such volumes requires high costs.

A significant reduction in salt pollution can be achieved by additional purification of mine waters with the help of aquatic plants. Among the aquatic plants studied by many scientists, the water hyacinth (Eichhornia) is considered the most effective plant for absorbing and splitting almost all chemicals into components.

In addition, Eichhornia has no equal in the speed of vegetation and the transformation of nitrogen, ammonia, potassium, calcium, magnesium, mineral salts, hydrogen sulfide, phosphorus and even poisons. Since this plant is floating, its reproduction, movement and removal from reservoirs is easy. During the growing season, the presence of pollutants provides the plant with nutrition and promotes its intensive growth (Krot 2006; Kravets et al. 1999). In addition, many useful products can be obtained by using the surpluses formed as a result of the rapid growth of euchronia (Gunnarsson and Petersen 2007; Jafari 2010).

10.3 Materials and Methods

In the Rostov region, similar studies were conducted on water treatment in small rivers. Eichhornia has been tested at all stages of water purification, but it turned out to be most effective at the stage of post-treatment in bio-settling ponds in the summer. The climate of the Rostov region is characterized by hot summers with an average temperature of +22 °C to +23° C in July and –5 °C in January, however, in some periods the air temperature can range from –20 °C to –25 °C. The duration of the frost-free period is 180–190 days. Such climatic conditions do not allow the use of water hyacinth in open reservoirs to purify polluted water throughout the year. That is, for almost half a year, polluted water will be discharged into rivers. However, urban wastewater and mine water pumped from the depths have huge reserves of heat.

Since there is no data on vegetation and preservation of water hyacinth in winter, the possibility of providing favorable conditions for viability and obtaining a large number of seedlings in winter is studied.

It is supposed to do this by using the mine water heat, which will allow applying biological post-treatment effectively for a longer period and thereby reduce anthropogenic pollution of small rivers. In total, there are three sections of underground water treatment facilities (TF) on the territory of the former mines: Mayskaya, Yuzhnaya and Glubokaya.

The volume of water pumping at the TF of Yuzhnaya mine is 450 m^3/hour, the water temperature ranges from 14.3 ± 0.2 °C, and at the TF of Mayskaya mine is 140 m^3/hour at a water temperature of 12.6 ± 1.4.

Compared to the mentioned TFS, the most favorable year-round temperature for plant conservation was observed at the TF Glubokaya mine, where water is pumped from a depth of 100 m. The pump capacity throughout the year is 930 m^3/hour, and the water temperature is 19.1 °C. After passing through the aerator, the sump and the first pond, the water even on the coldest days has a temperature of about 18.0 °C.

Thus, water hyacinth can be preserved in water due to the heat of mine wastewater, since the water temperature at the outlet of the treatment facilities did not fall below +7 °C. At the same time, deep autumn, winter and early spring are characterized by a large number of cloudy days and a short light period, which requires an assessment of the effect of illumination on the vegetative abilities of hyacinth.

To study temperature and humidity conditions, a digital thermohygrometer TGC-MG4.01 was used to monitor and record air temperature and humidity. When measuring the illumination created by incandescent lamps, fluorescent lamps and natural daylight, a TKA-LUX lux meter was used. Since the spectral composition of light also affects the vegetation of plants, the TKA-VD/02 spectrocolorimeter was applied to determine the correlated color temperature and the dominant wavelength.

10.4 Discussion of Research Results

The conducted studies and analyses of the thermal regime of the ponds show that in all settling ponds on the territory of the former Glubokaya mine in the winter period, even when the air temperature drops to –17 °C on some days, the temperature remains positive, i.e., the conditions may correspond to the necessary values for the conservation of Euchronia.

Previous observations show that the water hyacinth is very demanding of sunlight. The reduction of daylight hours and a large number of cloudy days in winter require studying the effect of illumination on the viability of plants.

To this end, a set of studies was carried out to determine the minimum illumination required for long-term storage of euchronia. Since the illumination even in an open space can change significantly in a short period, it was decided to take measurements in different periods—days, weeks, months. Based on the measurement results, minimum, maximum and average values were determined on clear and cloudy days.

The obtained average illumination values for the periods of the year in Shakhty are shown in Table 10.1.

Since it is impossible to provide the same lighting conditions in natural conditions, it was decided to estimate the minimum values of illumination in the laboratory by means of using artificial light sources.

For this purpose, studies were carried out by using the TKA-VD/02 spectrocolorimeter. The parameters were recorded by the device at a distance of 10 cm to 40 cm (10 cm interval) from the light source. Six lighting options were considered. The results of changing the illumination values E at different distances from the illuminator are shown in Table 10.2.

The highest values are observed for SPO-108 (LED) lamps, and the lowest values are observed for SQ0340-1503 and SAFFIT SBT1218 lamps, respectively. At this stage of the study, chromaticity diagrams, color temperature and wavelength indicators were obtained, which are presented in Table 10.3.

Table 10.1 Average illumination and chromaticity parameters in some periods of 2021 in Shakhty

Parameters	A clear winter day	A cloudy winter day	A clear summer day	A cloudy summer day	A cloudy autumn day
Illumination E, lux	7,600	698	35,747	16,276	6,873
Correlated color temperature T_c, K	5,886	7,412	5,782	6,104	6,234
Dominant wavelength l, nm	510	485	520	500	495

Table 10.2 Dependence of illumination E on the distance to the light source

Lamp brand	Type of lighting	Illumination E, lux distance from the lamp, cm			
		10	20	30	40
SPO-108 36 W	With a diffuser	624,090	445,540	235,200	175,880
	Without a diffuser	85,481	458,530	311,770	195,010
SQ0340-1503	НЈЛ-T8-20-G13	35,258	20,618	15,554	12,416
LDS 30 fluorescent	Without a diffuser	42,201	30,123	28,104	16,488
SAFFIT SBT1218	SAFX17081 18 W G13	28,576	24,074	14,575	10,750

Table 10.3 Color characteristics of different artificial light sources

Parameters	Lamp brands			
	SPO-108 18W	SQ0340-1503НЈЛ-T8-20	LDS 30 fluorescent	SAFFIT SBT1218 18 W G13
Correlated color temperature Tc, K	4,322	5,997	6,091	3,980
Dominant wavelength l, nm	575	500	510	575

As can be seen, the highest values of color temperature are marked for fluorescent lamp and the lamp SQ0340-1503 l, which correspond to cool shades of color. The lowest value has the Tc value of SAFFIT SBT1218 lamp, which corresponds to warm shades of color. The longest wavelength is 575 nm and corresponds to lamps SPO-108 and SAFFIT SBT1218. According to the results obtained, illumination close to natural is achieved at a distance of no more than 20 cm from the source.

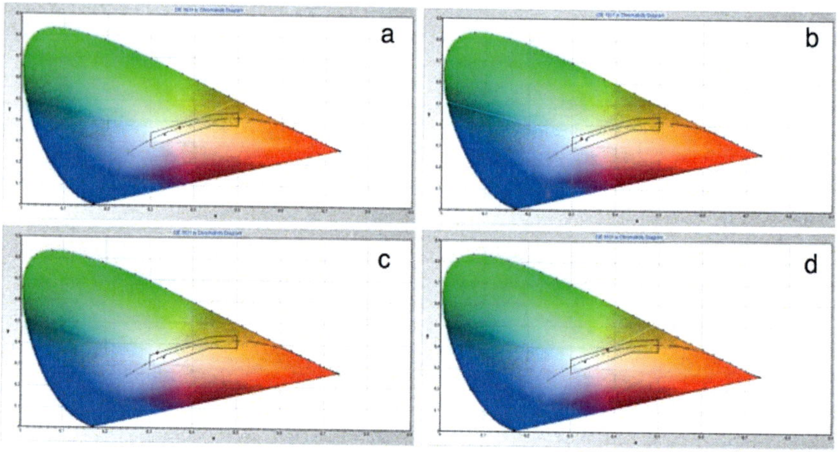

Fig. 10.1 Chromaticity diagrams of lamps in the CIE 1931 xy system

The chromaticity diagrams of the CIE 1931 xy system of the studied lamp types are shown in Fig. 10.1. Option "a" corresponds to the diagram obtained by measuring the illumination of the lamp SPO-108, option "b" to lamp SQ0340–1503, option "c" to fluorescent lamp LDS 30, option "d" to SAFFIT SBT1218.

An analysis of the chromaticity diagrams of various lighting options shows that the values of the color temperature of the lighting comparable to the first option (lamp SPO-108 without a diffuser) have practically not changed, approaching the mark of 4,298 K. The dominant wavelength l was fixed at 575 nm. When comparing the data obtained as a result of the study of two lighting options (SPO-108 lamps), it was found that the highest illumination values were without a diffuser.

When comparing with similar indicators of two variants of artificial lighting, it was revealed that on a clear day in the third decade of January and in the second decade of October, the illumination indicators are close to natural at a distance of 10 cm from the lamp without a light diffuser.

The minimum natural illumination was observed in 2020–2022 in the second half of December and included about 3,000 lx. This value is accepted for further research.

To assess the effect of the spectral composition of artificial lighting in the simulation, 4 different light sources were used, the characteristics of which are shown in Table 10.3. The required minimum illumination was achieved by installing two lamps at a distance of 21–26 cm from the surface of the sheet (Figs. 10.2 and 10.3).

For two months, the temperature of 17–18 °C and the illumination of 2,986–3,045 lx were in the plant cabinet. The lighting duration was 10 h. During this period, plant growth was not observed, even individual leaves died off, but in general, the plants looked healthy. After 2.5 months, the illumination was increased to 6,790 lx by reducing the distance to the light sources, at an air temperature of 24 °C and humidity of 67–70%, the duration of illumination remained the same.

Fig. 10.2 Illumination of Eichhornia seedlings with SQ0340-1503NL-T8-20 lamps

Fig. 10.3 Eichhornia seedlings illuminated by SAFFITSBT1218 lamps

This had the most favorable effect, the growth and formation of new processes resumed. However, the highest rate of vegetation was observed when the euchronia was illuminated by fluorescent lamps LDS 30 and phytolamps of the red-blue color SPECTRUM AFFITSBT1218.

In April 2023, the plants were transferred to an outdoor pool and are actively vegetating to this day.

10.5 Conclusions

According to the results of experimental studies in the town Shakhty, the following was determined: spectrum, color temperature and illumination in summer and winter.

As a result, of the study, it was found that the climatic features of the Rostov region and the level of illumination in the autumn–winter period do not correspond to the conditions of growth and vegetation of the higher aquatic plant Eichhornia crassipes due to the large number of cloudy days in winter. For preservation and better growth, additional light sources should be used.

Laboratory studies of various light sources allowed us to determine the most affordable and economical artificial light sources to provide illumination and color spectrum needed in the process of modeling the life of euchronia.

Bright lighting is necessary for the growth of euchronia. It has been experimentally established that in laboratory conditions, the most preferred light sources in the spectrum are LDS 30 fluorescent lamps and phytolamps of the red-blue spectrum. The optimal distance from the leaves to the light sources is 25–30 cm.

The minimum illumination of plants should be 3,000 lx with a duration of at least 10 h. The necessary optimal parameters for the conditions of rapid adaptation are the optimal temperature of 24 °C at a humidity of 65–70%.

The conducted research allows providing in laboratory conditions the necessary conditions for effective vegetation of euchronia in order to conduct an objective assessment of the effectiveness of wastewater treatment from various pollutants. This will allow us to develop technical and technological solutions for more efficient treatment of mine water in the conditions of the town Shakhty.

References

Agapov A, Navitni A, Kaplunov Y, Kharionovsky A (2007) Mine and quarry water of the coal industry: reference review. Central Publishing House, Moscow, p 357

Bespalova L, Sorokina V, Ivliev P (2006) The current state of the most important elements of the environmental situation in the Eastern Donbass and the trends of its change in connection with the liquidation of coal mines. In: Transactions of the Southern Scientific Centre of the Russian Academy of Sciences. Volume 1. Geology. SSC RAS Publishing-House, Rostov-on-Don, pp 164–180

Demkin VI, Kharionovsky A, Gusev N, Schastlivtsev E, Kaplunov V, Kirsch J (2009) On the problem of complex processing of mine waters. Mountain Inf Analyt Bull (Kuzbass-1) 7:209–219

Gunnarsson C, Petersen C (2007) Water hyacinths as a resource in agriculture and energy production: a literature review. Waste Manage 27:117–129

Jafari N (2010) Ecological and socio-economic utilization of water hyacinth (Eichhornia crassipes Mart Solms). J Appl Sci Environ Manag 14:43–49

Kravets V, Buhgalter L, Akolzin A, Buhgalter B (1999) Higher aquatic vegetation as an element of industrial wastewater treatment. Ecol Ind Russia 8:20–23

Krot Y (2006) Use of higher aquatic plants in biotechnologies of surface and wastewater treatment. Hydrobiological J 42(1):76–91

Kulikova A, Yu, S, Ovchinnikova T, Khabarova E (2020) Formation of mine waters and analysis of methods for their purification. Mining Inf Analyt Bull 7:135–145
Lashkov G (2011) Reserves and problems of fresh water in the world. [Resources and problems of fresh water in the world]. All-Russian scientific and technical conference [All-Russian scientific-technical conference]. Moscow, pp 306–309
Savenko V, Mikhaylov V, Zhuk V, Samokhin M, Zaslavskaya M, Frolova N (2012) Water reserves on earth [Resources of water on the earth]. Patterns of hydrological processes, pp 15–17
Shchadov V, Agapov A, Kaplunov Y, Navitni A (2003) Scientific and technical developments to protect water resources and wastewater treatment in the coal industry: review. Moscow, p 116
Smirnova A, Pozdnyakova N (2013) Fresh groundwater, the spread of aquifers and complexes [Fresh underground water, distribution of aquifers and complexes] Ecological and geographical atlas-book of the Voronezh region, pp 104–107
Yu, K, Limansky A, Gusev N (2011) Analysis of the problems of protection and rational use of water resources in terms of restructuring the coal industry. Monit Sci Technol 1(6):27–34

Chapter 11
Analysis of ASEAN Water Resource Policies in the Context of Belt and Road Initiative: A Perspective on Sustainable Development Goal 6 (SDG6)

Guansu Wang, Zhihong Huang, Dalin Li, and Jinyan Liao

Abstract Water scarcity is a pressing global challenge that has garnered increasing attention due to its multifaceted implications for ecosystems, human health, and socio-economic development. With the escalating influence of the Belt and Road Initiative (BRI), the connectivity among Southeast Asian Nations (ASEAN) countries has witnessed a pronounced intensification, particularly in the realm of water resources. The BRI's expansive infrastructure projects and economic collaborations have forged stronger ties within the ASEAN community, leading to increased cooperation and coordination on water-related issues. The study is grounded in the context of Sustainable Development Goal 6 (SDG6), which focuses on ensuring the availability and sustainable management of water and sanitation for all. By examining the intersection of SDG6 and ASEAN water resource policies, this research aims to reveal the current status, as well as potential opportunities and challenges, in achieving SDG6. The findings contribute to a deeper understanding of the intricate relationship between regional development initiatives and the sustainable management of water resources in ASEAN.

Keywords Belt and Road Initiative · Water resource policies · Association of Southeast Asian Nations · Sustainable Development Goal 6

G. Wang · J. Liao (✉)
Asia-Europe Institute, Universiti Malaya, Kuala Lumpur, Malaysia
e-mail: s2000152@siswa.um.edu.my

G. Wang
School of Humanities, Zhuhai College of Science and Technology, Zhuhai, Guangdong, China

Z. Huang
Zhuhai College of Jilin University, Zhuhai, Guangdong, China

D. Li
College of Computer Science and Technology, Jilin University, Changchun, Jilin, China

School of Computer Science, Zhuhai College of Science and Technology, Zhuhai, Guangdong, China

Z. Wan (ed.), *Water Resources Management and Water Pollution Control*, Environmental Science and Engineering, https://doi.org/10.1007/978-3-031-53456-0_11

11.1 Introduction

"Belt and Road Initiative" (BRI), introduced by China in 2013, stands as a crucial policy aimed at fostering economic cooperation along the historic Silk Road region (Qian et al. 2021). However, in the era of global interconnection and coordinated development, especially in Southeast Asia, the impact of the initiative has transcended traditional economic boundaries, extending its influence into critical realms such as water resources. The Association of Southeast Asian Nations (ASEAN), established in 1967, represents the oldest regional intergovernmental organization in Southeast Asia. Over the past two decades, the ten-countries ASEAN region has experienced changes such as urban expansion, intensive industrialization, and vigorous population growth (Xu et al. 2019). Geographically, the ten countries in ASEAN are mainly distributed in the Indochina region (Cambodia, Laos, Myanmar, Thailand, and Vietnam) and maritime areas (Brunei, Indonesia, Malaysia, Philippines and Singapore). In terms of ecological diversity, Southeast Asia mainly includes rainforests, rivers, lakes, and other important water resource types. Among them, Indonesian Borneo is the largest tropical rainforest except the Amazon River Basin in South America, and Malaysia's tropical rainforest is the largest on earth. One of the most biologically diverse ecosystems, it plays a pivotal role in the water resources system. In terms of river distribution, numerous rivers connect China with Southeast Asian countries, facilitating cooperation in water supply, agriculture, fisheries, energy production, ecosystems, transportation, and trade. The multifaceted collaboration in these domains underscores the intricate interplay between the "Belt and Road Initiative" and the water resources dynamics within the ASEAN region.

Yang and Li pointed out that BRI has an increasingly important impact on the development of water management in the ASEAN countries (Yang and Li 2019). First of all, the initiative explicitly outlines the construction of water conservancy infrastructure, encompassing irrigation systems, reservoirs, and canals. This infrastructure development holds the potential to substantially enhance agricultural production efficiency across ASEAN countries. Secondly, within the Belt and Road Initiative framework, collaborative efforts with select ASEAN countries involve the establishment of hydropower stations. This strategic partnership aims to generate clean energy through hydropower, addressing local power requirements while concurrently diminishing reliance on conventional energy sources. Thirdly, BRI provides a platform for ASEAN countries to establish a comprehensive water resources information sharing and monitoring system. This system facilitates real-time monitoring of crucial parameters such as water quality, quantity, and meteorological variations. Such data-driven insights contribute to an enhanced understanding of the current status of water resources within the region. Finally, the initiative catalyzes for promoting transnational water resources cooperation among ASEAN countries, particularly those sharing rivers, lakes, or groundwater. Under the impetus of the BRI, ASEAN countries increasingly prioritize inter-regional water resources cooperation. An illustrative example of such cooperation is observed among the Mekong River countries, which collaborate through mechanisms like the Mekong River Commission (MRC). Fei et al. found that this cooperative framework enables coordinated management of the Mekong River Basin's water resources (Fei et al. 2022). The undeniable impact of the BRICS initiative on the water resource policies of ASEAN countries demonstrates its role as a driving force in shaping the regional water resource management landscape.

Although the ASEAN countries have made progress in water management awareness and policies, the utilization and development of their water resources are still not optimistic. Firstly, according to ASEAN Secretariat (1995), anthropogenic activities in Southeast Asia have placed significant pressure on socioeconomic systems, and there are long-standing inequalities in clean water and sanitation in the distribution of water resources. It is estimated that up to 90% of the Wastewater is discharged directly into surface water resources such as lakes, rivers, reservoirs, estuaries, and coastal waters, thereby seriously polluting water resources (Connor 2015). Secondly, inadequacies in water resource management, coupled with factors like over-extraction of groundwater and the impacts of climate change, have led to water scarcity issues in some ASEAN countries, particularly about freshwater resources. In addition, some ASEAN countries adopt non-sustainable irrigation, and agricultural water occupies a large amount of water resources (Low 1991). For example, excessive extraction of groundwater and irrational agricultural drainage have led to soil salinization and water resource waste, exacerbating the severity of the problem. Furthermore, industrial emissions, agricultural pesticide use, urban drainage, and improper waste disposal contribute significantly to water pollution, exerting adverse effects on water quality. This intricate nexus of factors underscores the urgent need for addressing water resource sustainability in ASEAN countries within a comprehensive academic framework.

In addressing the water crisis, ASEAN countries have undertaken various initiatives. For instance, efforts have been made to establish a water quality assessment system aimed at standardizing wastewater treatment and water management practices (Novotny 2002) and proposed a blue economy framework (Hazra and Bhukta 2022).

However, there exists a developmental disparity in the water resource policies among ASEAN nations. Singapore, for instance, commenced its legislative efforts related to water pollution control and drainage as early as 1975. Over the ensuing years, Singapore has issued ten water resources management policies, demonstrating significant advancements in sewage treatment and public water management. This success is attributed to the continuous refinement and adjustment of policy content, coupled with the imposition of penalties to reinforce compliance. Other ASEAN countries, including Malaysia, Brunei, Cambodia, Indonesia, and the Philippines, have similarly enacted pertinent regulations on water resource management, establishing relatively stable management systems. In contrast, Vietnam, Myanmar, and Thailand have exhibited slower progress in the development of water resource policies. These countries only established their primary water resources laws in 2012, 2015, and 2018, respectively. In Laos, amidst the backdrop of the BRI and the globalization of water resources management, Laos has collaborated with China to construct a water resources testing and management platform, but Laos still has a difficult task in water resources management.

In addition to individual national water resource policies, the ASEAN countries face a critical need for and reliance on regional water resource cooperation policies. The United Nations, as part of its goal achievement perspective, regularly publishes comprehensive water resource policies for the 193 countries aligned with SDG goals. This involves the official release of data indicating the degree of management implementation on a scale from 0 to 100, thereby fostering the development and execution of water resource policies across nations. Studies indicate that government agencies worldwide take into consideration the Sustainable Development Goals (SDGs) when formulating water management strategies (Bridgewater 2021). Consequently, SDG6 assumes a pivotal role in shaping water resource management policies globally (Grafton et al. 2014).

While previous research has predominantly focused on the analysis of policy validity (Saimy and Yusof 2013) and the water quality index of specific ASEAN countries (Wong et al. 2020), it has lacked a relatively unified standard for evaluating regional water resource policies among ASEAN countries. Therefore, this article seeks to adopt a novel approach, starting from the standpoint of policy formulation. Utilizing text analysis methods, the study conducts an in-depth analysis of the water resource policies of the ASEAN countries. It employs the official and authoritative SDG6 indicators to scrutinize the water resource policies of these nations, to delineate their formulation and implementation within the SDG6 framework. The objective is to unveil the region's progress in achieving SDG6, highlight differences among countries, and explore prospects. Through this exploration, the article aspires to offer valuable insights into the dynamic interactions among ASEAN countries, fostering water cooperation to continually advance the Belt and Road Initiative and the pursuit of sustainable water resource management in ASEAN.

11.2 Materials and Methods

11.2.1 Sustainable Development Goal 6 (SDG6)

In the rapidly changing global landscape, and response to challenges such as income inequality, climate change, and gender equality, among others, the United Nations initiated the Sustainable Development Goals (SDGs) in 2015. The SDGs consist of 17 core objectives for global governments and businesses to collectively pursue sustainable development, serving as a crucial framework to address multifaceted challenges, including those related to water resources (UN General Assembly 2015). Among the 17 SDGs, Sustainable Development Goal 6 (SDG 6) stands out, aiming to ensure universal access to safe drinking water and sanitation for all. It places a particular emphasis on the sustainable management of water resources, wastewater, and ecosystems, recognizing the critical importance of favorable environmental conditions (Ho et al. 2020). SDG 6 was formulated against the backdrop of escalating global water-related crises, emphasizing the urgency of achieving universal access to safe drinking water, adequate sanitation facilities, and sustainable water resource management by 2030.

SDG6 has eight targets to be achieved by 2030. Progress toward the targets will be measured by using twelve indicators, which paint a comprehensive picture of water-related challenges and opportunities (Table 11.1). The target includes: Providing safe and affordable drinking water (target 6.1) and providing access to sanitation, and hygiene (target 6.2). And include improving water quality and wastewater treatment (target 6.3), increasing water-use efficiency and ensuring freshwater supplies (target 6.4), implementing IWRM (target 6.5), protecting and restoring water-related ecosystems (target 6.6). The two means of implementing these targets are to expand water and sanitation support to developing countries, and to support local engagement in water and sanitation management (target 6.a&b) (UN General Assembly 2020; Sadoff et al. 2020).

As nations globally strive to meet these targets, a nuanced examination of individual regions becomes imperative for a deeper understanding of the progress made and the unique challenges faced. Nurshafira analyzed that ASEAN countries possess abundant water resources, yet still face a series of water-related crises (Nurshafira 2018). Therefore, this study adopts this framework to analyze the performance of SDG6 in the ASEAN countries and compares them to the Southeast Asian region and globally (Fig. 11.1).

Firstly, overall, Southeast Asia's various indicators significantly outperform the global average indicators (except for Water Quality, Transboundary, and Cooperation, for which data from Southeast Asian countries are currently unavailable). This suggests that ASEAN countries have recognized the importance of sustainable water resource development and have implemented corresponding management measures. Secondly, at the country level, Singapore stands out with significantly higher scores across all indicators compared to other countries, particularly achieving 100% in Drinking Water, Sanitation, Wastewater, and Water Quality. Apart from Singapore, Malaysia also holds an advantageous position among the ASEAN members. However, Laos exhibits notably lower scores across various indicators, indicating a considerable disparity in awareness of sustainable water resource management among the countries. Additionally, when examining specific indicators, ASEAN countries show overall poor water resource utilization efficiency, significant variations in water scarcity levels, and Indonesia, Vietnam, and Cambodia receiving more sustainable development aid. Particularly noteworthy is the incomplete data for the Water Quality, Transboundary, and Participation indicators in ASEAN countries.

In conclusion, this illustrates an uneven distribution of water resources and imperfect water resource management in ASEAN countries. Therefore, SDG6 can serve as a valuable reference for water resource management in ASEAN countries, given its impact on livelihoods, socio-economic development, and ecosystem sustainability. Consequently, this study will focus on SDG6 indicators, analyzing the formulation and implementation of water resource policies in ASEAN countries within this framework.

Table 11.1 Goal 6. Ensure availability and sustainable management of water and sanitation for all (UN General Assembly 2015)

Goals and targets (From the 2030 Agenda for Sustainable Development)	Indicators
6.1 By 2030, achieve universal and equitable access to safe and affordable drinking water for all	6.1.1 Proportion of population using safely managed drinking water services
6.2 By 2030, achieve access to adequate and equitable sanitation and hygiene for all and end open defecation, paying special attention to the needs of women and girls and those in vulnerable situations	6.2.1 Proportion of population using (a) safely managed sanitation services and (b) a hand-washing facility with soap and water
6.3 By 2030, improve water quality by reducing pollution, eliminating dumping and minimizing release of hazardous chemicals and materials, halving the proportion of untreated wastewater and substantially increasing recycling and safe reuse globally	6.3.1 Proportion of domestic and industrial wastewater flows safely treated
	6.3.2 Proportion of bodies of water with good ambient water quality
6.4 By 2030, substantially increase water-use efficiency across all sectors and ensure sustainable withdrawals and supply of freshwater to address water scarcity and substantially reduce the number of people suffering from water scarcity	6.4.1 Change in water-use efficiency over time
	6.4.2 Level of water stress: freshwater withdrawal as a proportion of available freshwater resources
6.5 By 2030, implement integrated water resources management at all levels, including through transboundary cooperation as appropriate	6.5.1 Degree of integrated water resources management
	6.5.2 Proportion of transboundary basin area with an operational arrangement for water cooperation
6.6 By 2020, protect and restore water-related ecosystems, including mountains, forests, wetlands, rivers, aquifers and lakes	6.6.1 Change in the extent of water-related ecosystems over time
6.a By 2030, expand international cooperation and capacity-building support to developing countries in water- and sanitation-related activities and programmes, including water harvesting, desalination, water efficiency, wastewater treatment, recycling and reuse technologies	6.a.1 Amount of water- and sanitation-related official development assistance that is part of a government-coordinated spending plan
6.b Support and strengthen the participation of local communities in improving water and sanitation management	6.b.1 Proportion of local administrative units with established and operational policies and procedures for participation of local communities in water and sanitation management

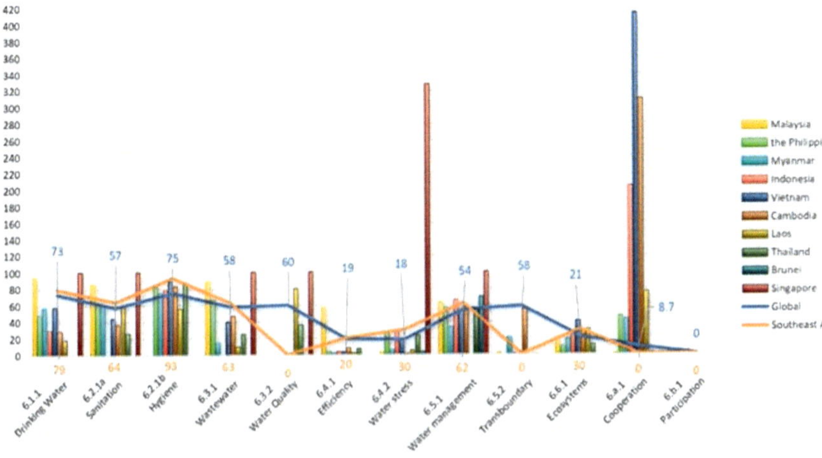

Fig. 11.1 Summary of SDG6 indicators in ASEAN countries

11.2.2 Methodology

Public policies are strategic plans crafted and adopted by authorities to address societal challenges and achieve public objectives in the collective interest of the community (Knickmeyer 2020). The analysis of policy texts related to water management offers valuable insights into the content and focus of a country's water resource management. Content analysis is a method that categorizes text content, such as deductive coding and word frequency statistics (Krippendorff 2018). It is a trending and effective method in the field of public policy (Hall and Steiner 2020). In contemporary approaches, Natural Language Processing (NLP) technologies (Bird et al. 2009) or Large Language Models (LLM) are increasingly employed to facilitate deductive coding processes (Chew et al. 2023). These technological advancements streamline tasks such as word frequency counting and text classification, particularly beneficial in research involving extensive text volumes, ensuring both efficiency and accuracy (Grimmer and Stewart 2013).

This study first utilized a pre-trained Large Language Model (LLM) for the text classification task (deductive coding) through the Zero-shot Text Classification method. Then, it used the Natural Language Toolkit (NLTK) Python library to count word frequencies based on the results from different countries and labels. Finally, an analysis of the water resource policies of the ASEAN countries was conducted. All these data processes were implemented using Python 3.11.

Table 11.2 An overview of processed data

Country	Preprocessed data counts	Country	Preprocessed data counts
Malaysia (MA)	1,069	Cambodia (CA)	659
Philippines (PH)	1,351	Laos (LA)	1,262
Myanmar (MY)	363	Thailand (TH)	1,268
Indonesia (IN)	1,773	Brunei (BR)	300
Vietnam (VI)	1,334	Singapore (SI)	1,474

11.2.3 Data Collection, Preprocessing and Cleaning

The data collection process for this study initially involves the identification and retrieval of relevant policy documents from the official government websites, publications, and authoritative databases of the ASEAN member countries, using keywords such as 'water management,' 'water policy,' 'water act,' among others. These documents encompass national water policies, strategic plans, and legislation related to water resource management, protection, and sustainability. To ensure the information is current and accurate, the data collection captures policy updates and revised versions, with policy texts meeting the defined criteria being downloaded for collection (UN General Assembly 2020). A total of 61 policies were collected. Among ASEAN countries, the amount of most countries ranges between 5 and 8. Singapore has the highest amount with 10 documents, while Myanmar has the fewest at 3. The earliest document dates to 1913 and the latest is from 2022.

A single policy document may include multiple sections related to various content. To ensure validation, this study employed 'full stops' as delineators to segment policy clauses, excluding data containing less than 5 words. An overview of the processed data is presented in Table 11.2.

For the preprocessed data, most countries have data counts ranging from approximately 1,000 to 1,400. Myanmar and Brunei each have only around 300 data points, while Indonesia has the highest count at 1,773.

Table 11.3 Codebook and definitions

Category	Definition
Drinking Water	Policy clause related to the usage of safe drinking water services
Sanitation	Policy clause related to the usage of safely managed sanitation services
Hygiene	Policy clause related to the usage of hand-washing facilities with soaps and water
Wastewater	Policy clause related to treatment of domestic and industrial wastewater flows
Water Quality	Policy clause related to bodies of water with good ambient water quality
Efficiency	Policy clause related to water-use efficiency
Water Stress	Policy clause related to freshwater withdrawal stress
Water Management	Policy clause related to integrated water resources management
Transboundary	Policy clause related to water-related transboundary basin area cooperation
Ecosystems	Policy clause related to water-related ecosystems
Cooperation	Policy clause related to water- and sanitation-related official development assistance
Participation	Policy clause related to local administrative units and procedures for participation of local communities in water and sanitation management

11.2.4 Codebook

The codebook of this study was constructed based on SDG6 from the Global Indicator Framework for the SDGs published by the United Nations (UN General Assembly 2020). The framework and its definitions provided in this study are presented in Table 11.3.

11.2.5 Data Analysis

Large Language Models (LLMs) represent a recent advancement in Natural Language Processing (NLP). BART, introduced by Facebook AI in 2019, is a transformer-based model specifically designed for natural language comprehension (Lewis et al. 2019). Models pre-trained on the Multi-genre Natural Language Inference (MNLI) datasets, such as BART-LARGE-MNLI, exhibit strong performance in tasks like zero-shot classification for short texts (deductive coding) (Yin et al. 2019). In this study, the bart-large-mnli model is utilized to categorize the text data with the 'topic_descriptions' parameter (definitions of each category) used for label definition. Some categories, including Hygiene, Efficiency, Water Stress, Transboundary, and Cooperation, showed lower accuracy rates. This is attributed to the limited relevance of these categories to policy texts and the constraints of the zero-shot classification approach. To ensure the reliability of this study, only the remaining 7 categories were selected for further analysis (Drinking Water, Sanitation, Wastewater, Water Quality, Water Management, Ecosystems, and Participation).

Word frequency analysis is a content analysis method that segments sentences into words and counts the frequency for further analysis. NLTK is a Python library commonly used for word frequency analysis in English text. It can process the data by removing stop words and stemming the words (Bird et al. 2009). In this study, NLTK is utilized to process text data, followed by counting the words and creating a word cloud.

11.3 Result and Discussion

11.3.1 Analyzing ASEAN Water Resource Policies: Key Themes and Regional Variances

The present water resources policies of ASEAN countries encapsulate information across various dimensions, spanning from resource allocation to environmental protection and socio-economic development. These policies mirror the strategies and priorities of these countries in water resources management. Analyzing the results of water resource policy encoding by country, utilizing non-linear color mapping by applying square root formulation (Fig. 11.2), reveals "water management" as the most frequently used term. The frequency index of this term in the water resources policies of all ASEAN countries surpasses 10, averaging 19.62. This indicates a gradual increase in attention to water resources management among ASEAN countries over recent decades. Additionally, the term "Participation" holds the second-highest frequency index, averaging 12.55. This underscores the pressing need for public involvement in water resources management across the ASEAN countries. Thus, based on the seven main indicators of SDG6, the current water resources policies of the ASEAN countries prioritize aspects such as water management, participation, water quality, and wastewater. Future policy formulation could consider strengthening the management of Drinking Water, Sanitation, and Ecosystems to more comprehensively meet the indicator requirements of SDG6.

ASEAN is currently faced with the challenge of safeguarding and efficiently utilizing water resources to meet present and future requirements. Consequently, they typically strike a balance among the three aspects of management mechanisms, natural resource distribution, and economic development when crafting water resources policies. The intricate interconnection between these elements becomes evident through the analysis of word frequency charts in 61 water resource policies across ASEAN countries.

In the assessment of water resources policies' implementation across diverse countries, it becomes apparent that the rights and responsibilities of the management mechanism significantly impact the efficiency of water resources management. In

Fig. 11.2 Water resource policy encoding results by country (non-linear color mapping)

Malaysia, upon reviewing the word frequency table and sorting out Malaysian policies, high-frequency terms predominantly include "commiss," "license," "approve," etc. (Fig. 11.3). Malaysia places a significant emphasis on management aspects in its water resources policy, particularly concerning issues of rights and responsibilities. The challenge of overlapping powers and responsibilities in water resources management within Malaysia is primarily manifested in the dynamics between the federal government and state governments, a situation stemming from the national system. Simultaneously, the frequency of "supply" and "service" in Malaysia's water resources policy also ranks 11th and 12th, indicating challenges related to decentralized water supply management and development. The fragmentation in supply and development within water resources management in Malaysia is a consequence of divergent state policies and uneven implementation (Saimy and Yusof 2013).

In contrast, the Philippines has swiftly realigned and redefined the distribution of powers and responsibilities in national water resources management, yielding significant positive outcomes. Analyzing the high-frequency word list in the Philippine water policy reveals key areas of focus, including "pollute," "management," "quality," "protect," etc. (Fig. 11.4). This emphasizes that the Philippine water resources policy centers around pollution control, quality testing, and safeguarding water resources.

An examination of policy developments in the Philippines highlights substantial advancements in water sources and decentralized laws between 2002 and 2010. These improvements can be attributed to the influence of strengthened water management institutions and river basin organizations, indicative of a systematic governance

Fig. 11.3 Word cloud diagram of Malaysia's water policies

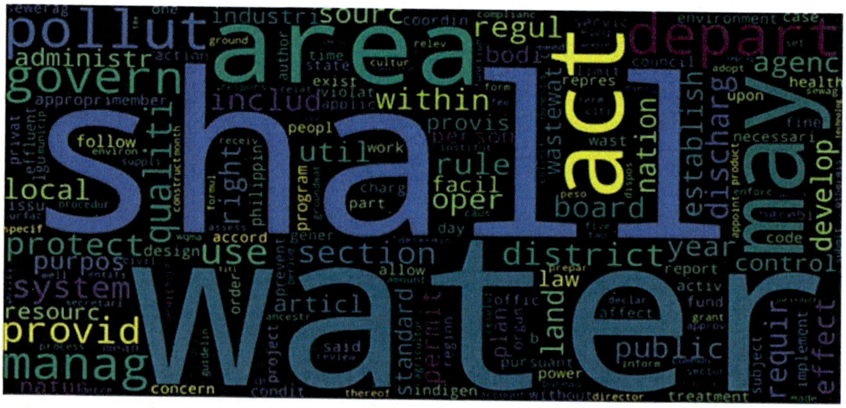

Fig. 11.4 Word cloud diagram of the Philippine's water policy

reform underway in the Philippines. The evolution of water policies, transitioning from the Marcos dictatorship to subsequent democratic phases, has been remarkable. Water management has progressively broadened, moving from a more centralized, unitary, policy-focused, and rights-based system to encompass watershed, pollution, and health considerations. The establishment of new governmental bodies demonstrates an increased emphasis on the uniform implementation of water resource policies, resulting in a marked improvement in management efficiency.

Secondly, while natural resources serve as the foundation of each country, the distribution characteristics of these resources also pose risks and challenges to water resources management. Myanmar serves as a typical representative. High-frequency words in the country's water resources policy mainly include "water", "resource",

Fig. 11.5 Word cloud diagram of Myanmar's water policy

"river", and "conserve" (Fig. 11.5), which are related to Myanmar's natural environment. Myanmar is rich in water resources, with a coastline extending for 2,832 km, a catchment area of ten major river basins of approximately 737,800 km^2, potential water resources of approximately 1,082 km^3 surface water, 495 cubic kilometers of groundwater, and an average annual rainfall of 7.99 feet (2,435 mm). However, this does not imply that Myanmar is devoid of water-related problems. The most prominent of which is the uneven distribution of rainfall, leading to floods, flash floods, water shortages, and droughts. In addition, due to the increase in urbanization and water demand in rural areas, and the expansion of economic activities such as agricultural irrigation, the pressure on the use of surface water and groundwater is gradually increasing. Therefore, how to effectively control and manage surface water and groundwater has become a key issue for the sustainable development of Myanmar's future water resources.

In addition, Vietnam is grappling with issues related to the uneven distribution of water resources and severe pollution in terms of natural resources. The country is home to numerous international rivers, with 60% of its water sources originating from neighboring countries. Consequently, any future increases in water usage by these neighboring countries may result in reduced water flows into Vietnam. Moreover, Vietnam's water source pollution problem is exacerbating due to the relentless pursuit of economic growth. Key terms such as "environment," "protect," "supply," and "waste" prominently feature in Vietnam's water resources management policies. Hence, ensuring the sustainable utilization of natural resources poses the primary challenge for Vietnam's current water resources policy.

Finally, the need for economic development drives the sustainable development of water resources management (Xu et al. 2019). According to the latest data from the Ministry of Tourism and Sports of Thailand, as of the first seven months of 2023, Thailand's total tourism revenue reached 1.08 trillion baht, with the number of international tourists entering Thailand exceeding 15 million, representing a year-on-year

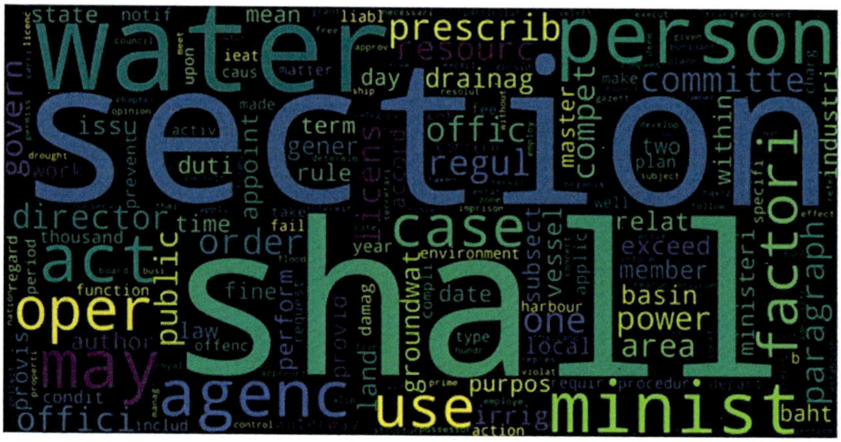

Fig. 11.6 Word cloud diagram of Thailand's water policy

increase of 384%. This highlights that the continuous development and expansion of tourism, reliant on water resources, has become a crucial economic pillar for Thailand. In Thailand's water policy, high-frequency words such as "development," "exceed," and "sustainable" (Fig. 11.6) underscore the focus on whether the relatively abundant water resources can sustain Thailand's long-term economic development. The new water law establishes the rights of state agencies to access local resources, but some NGOs and communities oppose water pricing as a means of taxing the poor. This further affirms that "water is competitive in all uses, has good economic value, and should be regarded as an economic commodity" (Dublin Principles).

11.3.2 Consistency Between ASEAN Water Resource Management and SDG6

From the research findings, it is evident that ASEAN countries exhibit consistency between their water resource management and SDG6. When formulating water resource policies, these countries typically consider how to align with the goals and indicators of SDG6 to ensure sustainable utilization of water resources, increase access to safe drinking water and sanitation facilities, and improve water quality.

Firstly, regarding indicators, the "Drinking Water" indicator for ASEAN countries, except for Malaysia (94) and Singapore (100), is below the global average score of 73. Similarly, the "Wastewater" indicator, except for Malaysia (89), the Philippines (67), and Singapore (100), is below the global average score of 68. This indicates that a lower proportion of the population in most ASEAN countries has access to safely managed drinking water services, and the proportion of safely treated wastewater is also low. This may reflect challenges in the quality of drinking water in some countries and issues with the effective enforcement of wastewater discharge standards, potentially influenced by factors such as water source pollution and inadequate treatment technologies. This aligns with our word cloud results (Figs. 11.7 and 11.8), where high-frequency words in ASEAN countries' policies related to drinking water and wastewater are broad and generic. For example, "suppli", "drink", "person", "service", "use", and "system", suggesting the current policies in these areas may not be comprehensive.

Conversely, the "Sanitation" indicator, except for Vietnam (44), Cambodia (37), and Thailand (26), is higher than the global average (59). The "Water Management" indicator for ASEAN countries, except for Myanmar (33), Vietnam (52), and Thailand (53), is higher than the global average score of 54. This suggests

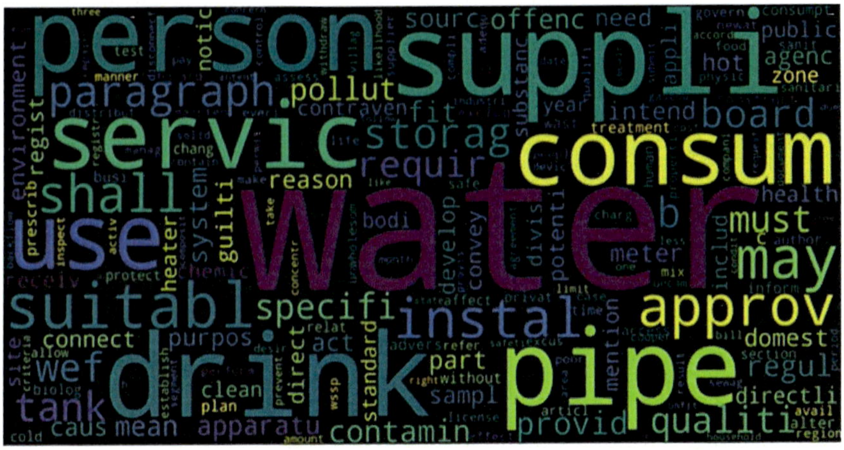

Fig. 11.7 Word cloud diagram of "drinking water"

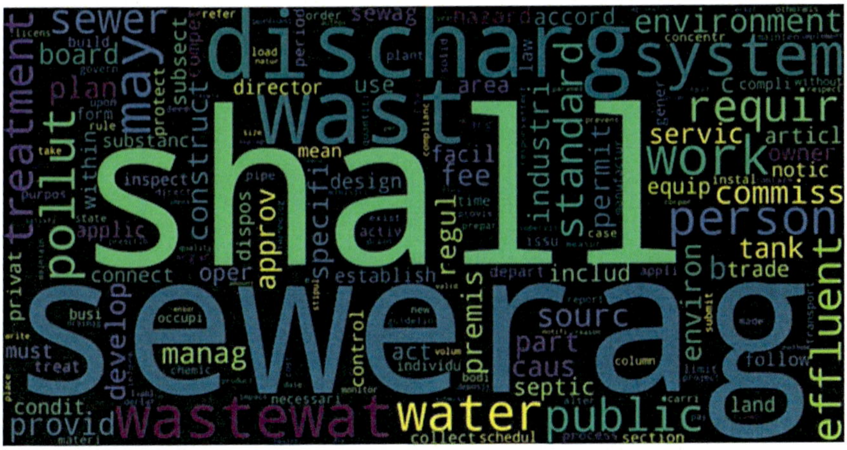

Fig. 11.8 Word cloud diagram of "wastewater"

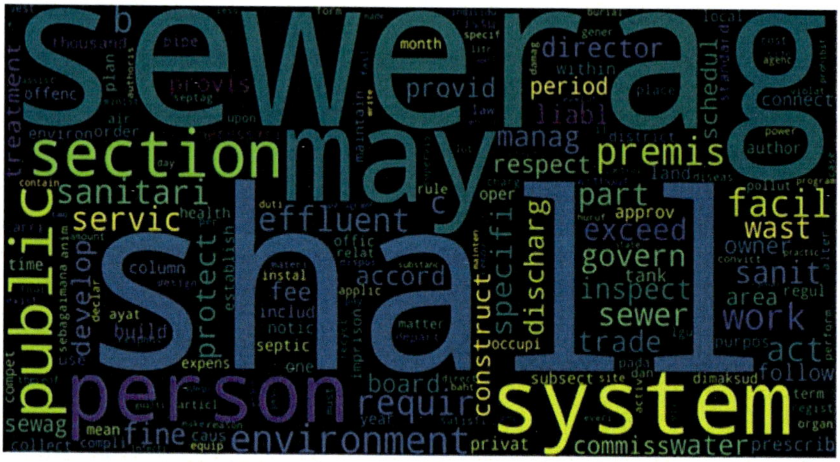

Fig. 11.9 Word cloud diagram of "sanitation"

that, compared to global reference standards, most ASEAN countries have a higher proportion of the population with access to environmental sanitation services, and the degree of execution in comprehensive water resource management is higher. This reflects some success in providing sanitation facilities, improving sanitary conditions, and adopting effective measures in water resource management and sustainable utilization, achieving a relatively good balance between water resource demand and protection. This aligns with our word cloud results (Figs. 11.9 and 11.10), where high-frequency words in ASEAN countries' policies related to sanitation and water management are detailed. For example, "sewerage", "effluent", "sewer", "plan", "area", and "system", which indicating relatively comprehensive policies in these areas.

Furthermore, at the national level, Singapore's SDG6 indicators are significantly higher than the global and other ASEAN countries, consistent with our research results. Looking at the word cloud (Fig. 11.11) high-frequency words in Singapore's water-related policies include terms like "require", "offend", "require", and "approve", reflecting a stringent approach. This includes the use of Mandatory Language to ensure the execution of specific actions or requirements, Enforcement Language for strong wording on dealing with violations, and formal language for approval and review. Singapore's success in water resource management is attributed to the formulation of long-term plans over the years, setting clear goals, timely implementation, and integration of water resource management into the overall development plans of the city-state (Lafforgue and Lenouvel 2015; Tortajada et al. 2013).

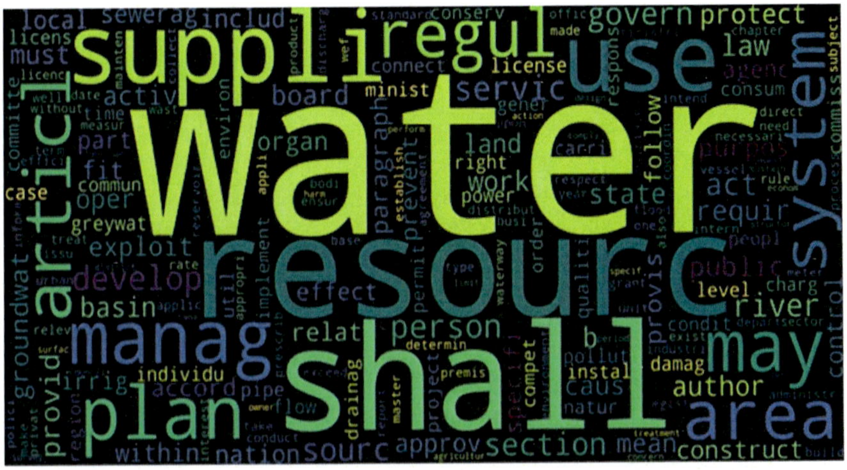

Fig. 11.10 Word cloud diagram of "water management"

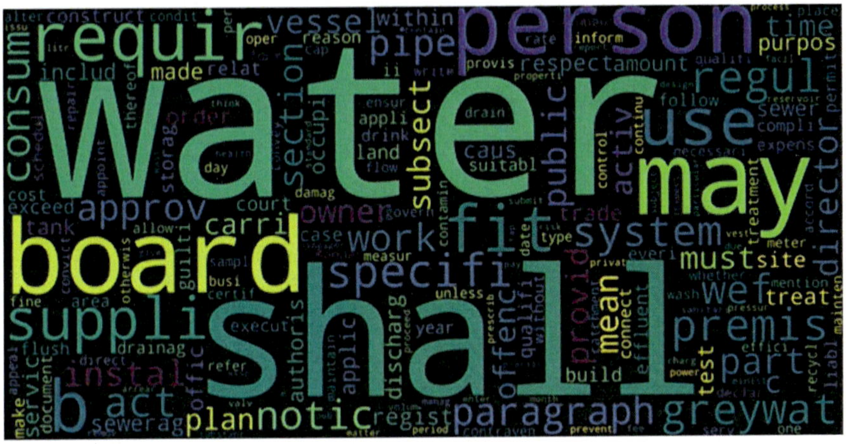

Fig. 11.11 Word cloud diagram of Singapore's water policy

In summary, there is a close relationship between the water resource policies of ASEAN's ten countries and SDG6, with both complementing each other. Governments, when managing water resources, can consider SDG6 as one of the references, with the potential to achieve more sustainable and inclusive water resource management.

11.4 Conclusion

The analysis of ASEAN water resource policies provides valuable insights into the key themes and regional variances shaping water management strategies across the member countries. The study reveals a predominant focus on "water management" and "participation," indicating a shared emphasis on effective resource utilization and public involvement. However, the research also highlights specific challenges and nuances in policy implementation, showcasing the intricate dynamics between management mechanisms, natural resource distribution, and economic development.

However, this study has certain limitations. The analysis primarily relies on policy documents, and a more comprehensive understanding could be achieved through stakeholder interviews and field studies. Additionally, the dynamic nature of water resource management necessitates continuous monitoring and adaptation. To address the identified challenges, future policy development should consider strengthening aspects such as drinking water, sanitation, and ecosystems, ensuring a more comprehensive coverage of SDG6 indicators. Cross-country collaboration and knowledge sharing within the ASEAN region can contribute to the development of standardized approaches, fostering a more unified and effective response to water management challenges.

In conclusion, as ASEAN countries continue to grapple with the complexities of water resource management, collaborative efforts, adaptive policies, and a commitment to sustainable development will be essential for ensuring a resilient and inclusive future for the region.

References

ASEAN Secretariat (1995) ASEAN: an overview

Bird S, Klein E, Loper E (2009) Natural language processing with Python: analyzing text with the natural language toolkit. O'Reilly Media

Bridgewater P (2021) A commentary on ecohydrology as a science-policy interface in implementing the UN Sustainable Development Goals. Ecohydrol Hydrobiol 21(3):387–392

Carbonell L, Hofmann P, Srikissoon N, Campos LC, Mbatha S, Lakhanpaul M, Mabeer V, Steenmans I, Parikh P (2023) Localisation of links between sanitation and the Sustainable Development Goals to inform municipal policy in eThekwini municipality, South Africa. World Dev Sustain 2:100038

Chew R, Bollenbacher J, Wenger M, Speer J, Kim A (2023) LLM-assisted content analysis: using large language models to support deductive coding. arXiv

Connor R (2015) WWAP: The United Nations world water development report 2015: water for a sustainable world. UNESCO

Fei YU, Hui-Juan CUI, Quan-Sheng GE (2022) Evaluation of water-related adaptation measures in nationally determined contributions of Belt and Road countries. Adv Clim Chang Res 18(1):70

Feldman DL (2022) Lecture notes in water policy (Vol. 11). World Scientific

Grafton QR, Wyrwoll P, White C, Allendes D (2014) Global water issues and insights. ANU Press

Grimmer J, Stewart BM (2013) Text as data: the promise and pitfalls of automatic content analysis methods for political texts. Polit Anal 21(3):267–297

Hall DM, Steiner R (2020) Policy content analysis: qualitative method for analyzing sub-national insect pollinator legislation. MethodsX 7:100787
Hazra S, Bhukta A (2022) The blue economy: an Asian perspective. Springer Nature
Ho L, Alonso A, Forio MAE, Vanclooster M, Goethals PL (2020) Water research in support of the Sustainable Development Goal 6: A case study in Belgium. J Clean Prod 277:124082
Knickmeyer D (2020) Social factors influencing household waste separation: a literature review on good practices to improve the recycling performance of urban areas. J Clean Prod 245:118605
Krippendorff K (2018) Content analysis: an introduction to its methodology. Sage Publications
Lafforgue M, Lenouvel V (2015) Closing the urban water loop: lessons from Singapore and Windhoek. Environ Sci: Water Res Technol 1(5):622–631
Lewis M, Liu Y, Goyal N, Ghazvininejad M, Mohamed A, Levy O, Stoyanov V, Zettlemoyer L (2019) BART: denoising sequence-to-sequence pre-training for natural language generation, translation, and comprehension. arXiv
Low KS (1991) Ecology of cities in the humid tropics: water-related issues
Morris JC (2022) Clean water policy and state choice: promise and performance in the water quality act. Cambridge University Press
Novotny V (2002) Water quality: diffuse pollution and watershed management. John Wiley & Sons
Nurshafira T (2018) Assessing the origin of ASEAN Water Resources Management Cooperation: The future of ASEAN values. In: 50 years of amity and enmity: the politics of ASEAN Coorperation, 1, p 199.
Potivejkul S, Pimdee P (2018) Water ozonisation/ozonation for ASEAN environmental resource sustainability: a structural equation model analysis. Asia Pac Soc Sci Rev 17(3):142–154
Qian X, Bai Y, Huang W, Dai J, Li X, Wang Y (2021) Fuzzy technique application in selecting photovoltaic energy and solar thermal energy production in belt and road countries. J Energy Storage 41:102865
Sadoff CW, Borgomeo E, Uhlenbrook S (2020) Rethinking water for SDG 6. Nat Sustain 3(5):346–347
Sabrina D, Azzahra, AU (2023) Visi air ASEAN dalam pengaturan pengelolan sumber daya air terpadu di Indonesia. Jurist-Diction 6(2)
Saimy IS, Yusof NAM (2013) The need for better water policy and governance in Malaysia. Procedia Soc Behav Sci 81:371–375
Tortajada C, Biswas AK (2013) Asian perspectives on water policy. Routledge
Tortajada C, Joshi YK, Biswas AK (2013) The Singapore water story: sustainable development in an urban city-state. Routledge
UN General Assembly (2015) Transforming our world: the 2030 agenda for sustainable development
UN General Assembly (2020) Global indicator framework for the Sustainable Development Goals and targets of the 2030 Agenda for Sustainable Development
Wong YJ, Shimizu Y, He K, Nik Sulaiman NM (2020) Comparison among different ASEAN water quality indices for the assessment of the spatial variation of surface water quality in the Selangor River basin Malaysia. Environ Monit Assess 192(10):644
Xu R, Chou LC, Zhang WH (2019) The effect of CO_2 emissions and economic performance on hydrogen-based renewable production in 35 European countries. Int J Hydrogen Energy 44(56):29418–29425
Yang Y, Li F (2019) The Belt and Road Initiative: ASEAN countries' perspectives (Vol. 8). World Scientific
Yin W, Hay J, Roth D (2019) Benchmarking zero-shot text classification: datasets, evaluation and entailment approach. arXiv

Chapter 12
Analysis of Water Pollution Situation and Measures in Guxian Street of Liyang City

Wei Tang, Yangyang Tang, Zhenhong Zhu, and Dechao Chen

Abstract In line with Xi Jinping's ecological civilization vision and the principle that regards "green mountains as valuable as gold and silver," the Liyang Municipal People's Government has prioritized comprehensive rectification efforts for the ecological and water environment in Guxian Street. Implementing a strategic approach, a series of measures has been systematically introduced to elevate the water environment conditions in Guxian Street, resulting in significant advancements in recent years. Aligned with the ecological civilization construction directives issued by national, provincial, and municipal authorities, this study seeks to provide scholarly guidance for the holistic management of the water environment in Guxian Street. The primary objective is to achieve a state of stable water quality within the region. This concerted effort not only enhances the local environment but also aims to lay a solid foundation for Liyang City to be acknowledged as a "Yangtze River Delta Eco-Innovation Demonstration City." The multifaceted approach adopted by the Liyang Municipal People's Government underscores its commitment to balancing ecological preservation with sustainable development.

Keywords Water pollution · Analysis · Measures

W. Tang · Z. Zhu · D. Chen (✉)
School of Geography Science and Geomatics Engineering, Suzhou University of Science and Technology, Suzhou, Jiangsu 215009, People's Republic of China
e-mail: dcchen2002@163.com

Y. Tang
Jiangsu Shike Environmental Development Co. Ltd, Suzhou, Jiangsu 215009, People's Republic of China

Z. Wan (ed.), *Water Resources Management and Water Pollution Control*,
Environmental Science and Engineering, https://doi.org/10.1007/978-3-031-53456-0_12

12.1 Materials and Methods

12.1.1 Overview of the Study Area

Guxian Street, situated in Liyang City, Jiangsu Province, holds prominence as the "Pearl of Jiangnan" and the "back garden of the city" within China's Yangtze River Delta. Positioned at the geometric center of Suzhou, Zhejiang, and Anhui Province, it falls within the Yangtze River Delta regional planning framework characterized by the "one nucleus and nine axes" model along the Shanghai-Nanjing corridor. As an integral focal point in the development strategy of the national Ning-Hangzhou ecological economic belt, Guxian Street also contributes significantly to the cross-radiation zone along the Shanghai-Nanjing and Ning-Hangzhou development axes, as stipulated in the "one core and nine axes" configuration of the Yangtze River Delta regional planning. Belonging to the Taihu Lake water system, Guxian Street channels all its water into Taihu Lake, with key transit rivers including the Chating River and Li Dai River. This region experiences a northern subtropical monsoon climate characterized by distinct seasons–dry, wet, cold, and warm. It boasts abundant rainfall, ample sunshine, an extended frost-free period, and relatively abundant temperature and water resources. The monsoonal influence results in unevenly distributed rainfall in both temporal and spatial dimensions. Based on Liyang meteorological station observations, the average annual temperature stands at 16.8 °C, average annual precipitation records 1,193.9 mm, average annual relative humidity is 75%, with easterly winds prevailing throughout the years and an average annual wind speed of 1.9 m/s (Wang and Duan 2018).

12.1.2 Pollutant Flux Calculation Methods

Generic formula for calculating the flux of a pollutant over a period of time (Zhang et al. 2021):

$$W = k \int C(t) * Q(t) * dt \quad (12.1)$$

where: W is the pollutant flux, C (t) is the pollutant concentration at time t (unit mg/L); Q (t) is the flow rate at time t (unit m/s), and k is the unit conversion factor.

12.2 Forms and Situations Faced

12.2.1 Water Resources Situation

In 2021, industrial enterprises within Guxian Street collectively utilized approximately 555,175,000 cubic meters of freshwater. The predominant water consumer in this locale is Jiangsu Top Plastic Industrial Co. Ltd., contributing significantly to the overall water usage. Specifically, the freshwater consumption of Jiangsu Top Plastic Industrial Co. Ltd., amounted to approximately 52,416,000 cubic meters in 2021, representing 9.44% of the total freshwater consumption by industrial enterprises in the region. Moreover, regional fresh water use per unit of output 0.652 cubic meters/million yuan (Li et al. 2023).

12.2.2 Water Pollution Analysis

12.2.2.1 Industrial Waste Water

Drainage in the area adopts the rainwater and sewage diversion system, and rainwater is discharged into the river nearby. The sewage pipe network in the built-up area of Guxian Street has achieved full coverage. Regional sewage is connected to the Garden Sewage Treatment Plant for centralised treatment (Chen et al. 2023). As can be seen from Fig. 12.1, the key water pollution enterprises in Guxian Street are Buhler Machinery, Pengcheng Steel Structure, Jinlibao, etc., with annual wastewater emissions of 57,600 tonnes, 22,600 tonnes and 14,400 tonnes respectively; the regional unit of industrial output value of wastewater emissions of 0.288 tonnes/million yuan, and the unit of industrial output value of COD emissions of about 0.011 kg/million yuan, and the control of regional water pollution emissions is good.

12.2.2.2 Domestic Sewage

Domestic sewage is divided into urban domestic sewage and rural domestic sewage. Domestic pollution sources are determined with reference to the <Manual of Methods and Coefficients for Accounting for Production and Discharge in Statistical Surveys of Emission Sources-Manual of Coefficients for Accounting for Production and Discharge of Domestic Sources>.

1. Urban domestic sewage

There is one sewage plant in the district that collects sewage from towns and cities, namely, the Garden Sewage Treatment Plant, with a daily treatment scale of 30,000 m^3/d, which can accept sewage from towns and cities in the district.

2. Rural domestic sources of pollution

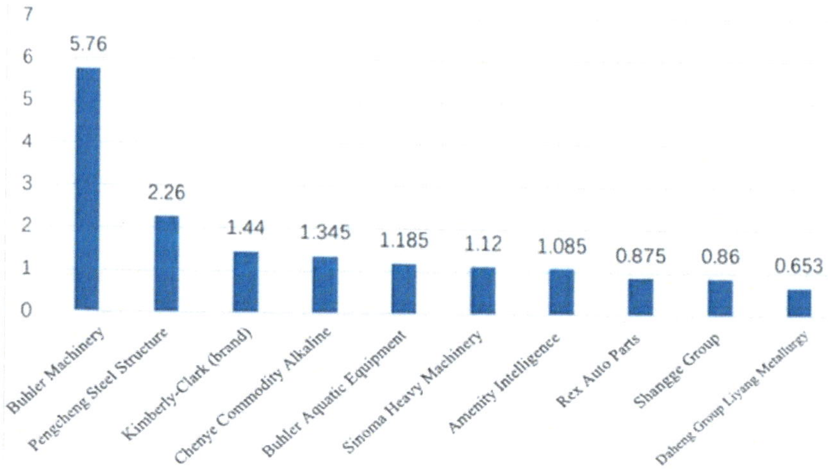

Fig. 12.1 Distribution of major wastewater pollution sources in Gu county streets (Unit: 10,000 tonnes)

In recent years, the municipal government of Liyang has actively advanced the rural domestic sewage remediation project. Presently, two primary approaches are employed for treating rural domestic sewage within the region. The first involves connecting to the sewage network and establishing centralized domestic sewage treatment facilities. Through this method, sewage is channeled to the designated sewage network and subsequently directed to the Garden Sewage Treatment Plant for centralized treatment. The second approach entails connecting to the official sewage website, where effluent from decentralized small-scale domestic sewage treatment facilities in rural areas is treated. Notably, the effluent from these facilities generally aligns with the Class A standard stipulated in the "Pollutant Emission Standards for Urban Sewage Treatment Plants" (GB18918-2002).

12.2.2.3 Agricultural Land Source

Agricultural surface sources within the Gouxian Street District encompass primarily farmland, garden land, and woodland. As of the conclusion of 2021, the district comprises approximately 30,002.4 acres of farmland, 11,627.1 acres of forested land, and 7,755.3 acres of parkland. These areas account for 32.29%, 12.51%, and 8.35%, respectively, of the total land area within the street (Fig. 12.2).

According to statistical data, Guxian Street witnesses an application of 548.7 tonnes (pure) of agricultural fertilizer and 5.03 tonnes (pure) of pesticides. The corresponding intensity of agricultural fertilizer application stands at 218 kg (pure) per hectare, while the intensity of pesticide application is 2 kg (pure) per hectare. The quantification of pollutants generated and discharged (lost) from agricultural sources adheres to the methodologies outlined in the "Manual on Production and

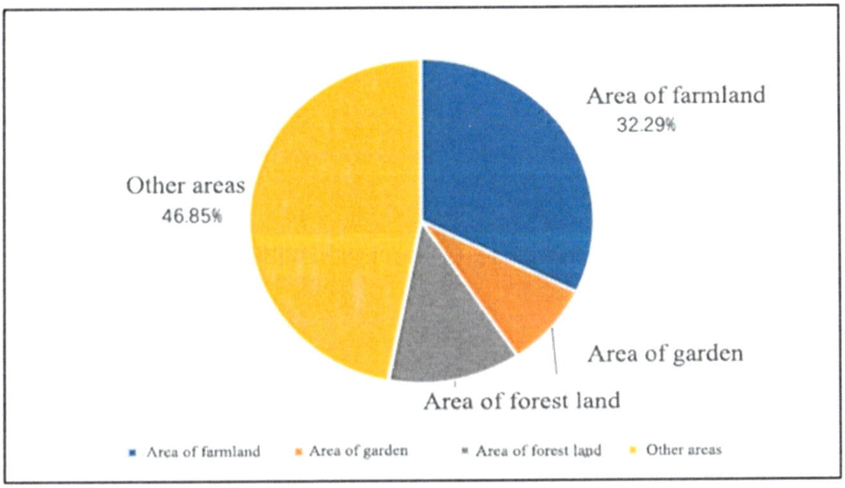

Fig. 12.2 Percentage of land use distribution on ancient county streets

Emission Accounting Methods and Coefficients for Statistical Surveys of Emission Sources—Manual on Production and Emission Accounting Coefficients for Agricultural Sources." Specifically, agricultural pollution sources contribute 2.024 tonnes/year of ammonia nitrogen, 16.422 tonnes/year of total nitrogen, and 1.491 tonnes/year of total phosphorus. As depicted in Fig. 12.3, the primary contributors to ammonia nitrogen, total nitrogen, and total phosphorus originate from agricultural land, with agricultural land exhibiting the highest proportion of total phosphorus emissions at 94.09%. Meanwhile, the contributions of ammonia nitrogen and total nitrogen from agricultural land are relatively comparable, accounting for 27.92% and 78.97%, respectively. Gardens, in contrast, exhibit a comparatively smaller share in the emissions of these three pollutant sources.

12.2.2.4 Analysis of the Current Status of Surface Water Quality

Guxian Street encompasses a total of four township assessment sections. From 2020 to 2021, the water quality standard compliance rate within these sections has consistently stood at 50%. Certain sections within the area continue to face challenges in maintaining stable adherence to the standards. Notably, areas such as Chating River Chating Bridge, Guxian River Wangxian Bridge, and Binhu Bridge frequently experience difficulties in achieving Class III water quality standards.

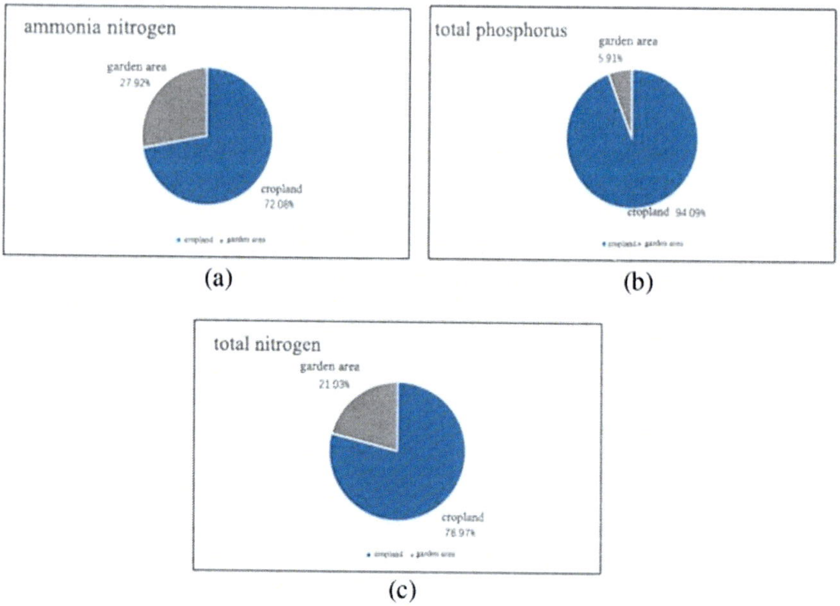

Fig. 12.3 Distribution of agricultural source pollutants (Ammonia nitrogen, total phosphorus, total nitrogen) as a percentage of emissions

12.3 Main Measures

12.3.1 Pollution Control Project

The sewage treatment system in the district includes a centralised system (Liyang City Garden Sewage Treatment System) and a decentralised system (Xu et al. 2022).

12.3.1.1 Centralised Sewerage System

Wastewater management in Guxian Street involves centralized disposal at the Liyang City Garden Sewage Plant. The initial phase of the plant boasts a capacity of 30,000 m^3/d, with plans underway for a second-phase expansion to reach 80,000 m^3/d. The plant's service coverage extends to the domestic sewage originating from the region east of Nanmen Road in Li Cheng Street, south of Nanhe River, Guxian Street, as well as Tianmu Lake Town and Dai Bu Town.

The Liyang City Garden Sewage Plant currently operates with a project capacity of 3.0 million m^3/d. Primarily designed for the collection and treatment of domestic sewage from Li City Street, the southern part of Guxian Street (encompassing the area east of South Street, south of Chengzhong River, and north of Yanshan River), Yanshan Area (covering the region south of Yanshan River and north of Yancheng

Avenue), as well as domestic sewage from Tianmu Lake Industrial Park, Tianmu Lake Town Township, and Dai Bu Town Township. The plant ensures that treated effluents meet the standards outlined in the "Environmental Quality Standards for Surface Water" (GB 3838-2002), specifically adhering to Table 12.1, Class IV standard limits for tail water COD, ammonia nitrogen, and total phosphorus. Additionally, other pollutants are treated to align with the "Taihu Lake Area Urban Sewage Treatment Plants and Key Industrial Industries Major Water Body Pollutant Emission Limits" (DB 32/1072-018), as detailed in Table 12.1 for Taihu Lake Basin Level 1 and Level 2 Protection Zone major water body pollutant emission limits. The discharged effluents are directed towards the South River (with a subsequent proposal for altering the outfall and changing the discharge route from the current status quo to the Old Dai Bu River).

The effluents from the Liyang City Garden Sewage Plant conform to the Surface Water Environmental Quality Standards (GB3838-2002), meeting the criteria set for Class IV standards. Total Nitrogen (TN) adheres to the "Taihu Lake area urban sewage treatment plants and key industrial sectors of the main water pollutants discharge limits" (DB 32/1072-2018), as specified in Table 12.1 for Taihu Lake Basin I and II

Table 12.1 List of major fresh water users in good county streets

Serial number	Company identification	Fresh water consumption (cubic metres)	Industrial output (million yuan)	Fresh water consumption per unit of output value (cubic metres per million yuan)
1	Jiangsu Top Plastic Industry Co.	52,416	13,445.4	3.898
2	Liyang Jinlibao Adhesive Products Co.	51,125	31,762.4	1.61
3	Jiangsu Ansui Intelligent Power Transmission Engineering Technology Co.	48,176	43,547	1.106
4	Liyang Lux Auto Parts Manufacturing Co.	41,227	49,892.9	0.826
5	Liyang Sinoma Heavy Machinery Co.	35,802	67,325	0.532
6	Jinyiyuan (Jiangsu) New Material Co.	32,863.34	10,774.6	3.05
7	Jiangsu Tianmu Construction Group Outline Machine Co.	30,333	12,348.1	2.456
8	Buhler (Changzhou) Machinery Co.	29,775	196,382.5	0.152
9	Shanghai Gear Group Co.	28,964	17,306.8	1.674
10	Jiangsu Pengcheng Steel Structure Group Co.	23,921	71,403.3	0.335

standards. Suspended Solids (SS) meet the "municipal wastewater treatment plant pollutant discharge standards" (GB18918-2002) at a Class A level and are subsequently discharged into the Lao Dai Bu River. Following the tail water treatment's compliance with standards, an additional purification step takes place through artificial wetlands. Out of the treated effluent, 20,000 m^3/d (25%) is earmarked for reuse, contributing to the replenishment of water in front of Yanhu Lake and Marriott Hotel. The remaining 60,000 m^3/d (75%) is directed towards discharge into the Old Dai Bu River (Table 12.2).

Surface Water Environmental Quality Standards (GB3838-2002) in Class IV standard; TN implementation of the "Taihu Lake area urban sewage treatment plants and key industrial sectors of the main water pollutants discharge limits" (DB 32/1072-2018) in Table 12.1 Taihu Lake Basin I and II standards; SS implementation of the "municipal wastewater treatment plant pollutant discharge standards" (GB18918-2002) a level of A standard) and then discharged to Lao Dai Bu River. After the tail water treatment meets the standard, it is further purified by artificial wetland, and 20,000 m^3/d (25%) of the effluent water is reused to replenish the water in front of Yanhu Lake and Marriott Hotel, and the remaining 60,000 m^3/d (75%) is discharged into the Old Dai Bu River.

At present, the Liyang City Garden Sewage Treatment Plant operates at an average daily capacity of approximately 26,700 m^3/d, representing over 80% of the designed capacity for the first phase, which is 30,000 m^3/d. During peak periods, the sewage treatment plant functions at full operational capacity. The second phase of the project is currently in the promotion stage. The sewage pipe network has been established beneath key thoroughfares such as South Street, Nanshan Avenue, and Yanshan Road, with a sewage main network ranging from d500 to d1000. For other roads, sewage pipes ranging from D400 to D500 have been laid based on specific requirements. The regional sewage is efficiently conveyed through the operational 7# sewage pumping station and the Nanshimen Road pumping station, ultimately reaching the Liyang City Garden Sewage Treatment Plant for centralized treatment.

12.3.1.2 Decentralised Treatment System

Twenty-six centralized domestic sewage treatment facilities have been established in the district, encompassing seven administrative villages. These facilities employ the "A/O contact oxidation + artificial wetland" treatment process, operating under a relatively centralized treatment mode. The effluent from the treatment facilities complies with the primary A standard stipulated in the "Pollutant Emission Standards for Urban Sewage Treatment Plants" (GB18918-2002). In 2021, the construction and transformation of environmentally friendly rural latrines were successfully completed, achieving a 100% penetration rate of harmless sanitary latrines in rural areas. The coverage of domestic sewage treatment facilities in administrative villages has been fully realized. Additionally, the construction and transformation of environmentally friendly rural household toilets will be accomplished in 2021, achieving a 100% penetration rate of harmless rural sanitary latrines. Full coverage of sewage

Table 12.2 Basic information of rural domestic wastewater treatment facilities in Guxian Street

Serial number	Administrative village	Natural village	Population (persons)	Processing mode	Treatment process
1	Ancient county	Dajang village	329	Relative centralisation	A/O contact oxidation + artificial wetland
2		Shengzhuang	323	Relative centralisation	A/O contact oxidation + artificial wetland
3	Tea pavilion	Fangji	478	Relative centralisation	A/O contact oxidation + artificial wetland
4		Xiaolintou	96	Relative centralisation	A/O contact oxidation + artificial wetland
5	Daling village	Hou village	390	Relative centralisation	A/O contact oxidation + artificial wetland
6		Yintang village	261	Relative centralisation	A/O contact oxidation + artificial wetland
7		Xujia village	480	Relative centralisation	A/O contact oxidation + artificial wetland
8	Xinlian village	Xincunli village	1,320	Relative centralisation	A/O contact oxidation + artificial wetland
9		Dafotang village	384	Relative centralisation	A/O contact oxidation + artificial wetland
10		Tang village	262	Relative centralisation	A/O contact oxidation + artificial wetland
11		Dongxu village	896	Relative centralisation	A/O contact oxidation + artificial wetland

(continued)

Table 12.2 (continued)

Serial number	Administrative village	Natural village	Population (persons)	Processing mode	Treatment process
12	Shanggelou village	Tangxia village	771	Relative centralisation	A/O contact oxidation + artificial wetland
13		Tangjia village	159	Relative centralisation	A/O contact oxidation + artificial wetland
14		Maojia village	144	Relative centralisation	A/O contact oxidation + artificial wetland
15		Shanggelou village	1,614	Relative centralisation	A/O contact oxidation + artificial wetland
16	Baijiatang village	Baijiatang	188	Relative centralisation	A/O contact oxidation + artificial wetland
17		Nanxishui	449	Relative centralisation	A/O contact oxidation + artificial wetland
18		Xiadayu	194	Relative centralisation	A/O contact oxidation + artificial wetland
19		Yuetan	300	Relative centralisation	A/O contact oxidation + artificial wetland
20		Tiansheqiao	150	Relative centralisation	A/O contact oxidation + artificial wetland
21	Xinqiao village	Xinqiao	101	Relative centralisation	A/O contact oxidation + artificial wetland
22		Yanxiangli	430	Relative centralisation	A/O contact oxidation + artificial wetland

(continued)

Table 12.2 (continued)

Serial number	Administrative village	Natural village	Population (persons)	Processing mode	Treatment process
23		Tianshe	78	Relative centralisation	A/O contact oxidation + artificial wetland
24		Qiaonanqiaotou	264	Relative centralisation	A/O contact oxidation + artificial wetland
25		Dongding	219	Relative centralisation	A/O contact oxidation + artificial wetland
26		Tangnan	207	Relative centralisation	A/O contact oxidation + artificial wetland
Add up			10,487		

treatment facilities in administrative villages will be attained concurrently. Furthermore, a dedicated rectification initiative, known as the "double 60 standards," will be simultaneously executed for the built facilities to ensure that their normal operation rate surpasses 80%.

12.3.2 *Deepening the Battle for Blue Water*

12.3.2.1 Fighting the Water Quality Assessment Section to Meet the Standard Rectification War

The upcoming initiatives involve a retrospective remediation plan that will meticulously assess sections where water quality may not consistently meet standards in 2021. This comprehensive strategy will encompass measures to control pollution sources, intercept pollutants, undertake dredging and water cleanup, initiate ecological restoration efforts, and implement emergency safeguards during flood seasons. Furthermore, there will be a concerted effort to reinforce industrial pollution control measures, compelling enterprises in pivotal industries to consistently adhere to sewage discharge standards (Liu et al. 2020).

12.3.2.2 Fighting the Battle for Comprehensive Remediation of the Taihu Lake Basin

Facilitate the proactive advancement of the sewage treatment demonstration area pilot construction (Zhang et al. 2015). Undertake a holistic approach to managing industrial, domestic, and agricultural surface pollution. Vigorously implement rain and sewage diversion, standardize river outfalls, and adhere to the entire process of sewage discharge standardized management. Additionally, display the rain and sewage network map publicly on the wall for enhanced transparency and awareness.

12.3.2.3 Continuing to Fight the Battle Against Black-Smelling Water Bodies

Persisting in endeavors to eliminate tributary streams and branches' shortcomings through the formulation and execution of a "one river, one policy" remediation plan. Intensifying the development of the sewage collection pipeline network through comprehensive drainage network censuses, repairs, and transformations (Xue et al. 2023). Advancing the rain and sewage diversion transformation in urban areas to augment the efficacy of urban sewage collection. Executing the Tianmu Xingcheng Plate old city renewal project, scrutinizing the river-adjacent pipeline network, and refurbishing the rain and sewage pipe networks in the old district and beneath roads, thereby ensuring a complete separation of rain and sewage.

12.3.2.4 Strengthening Synergistic Management of Pollution in Land-Based Waters

Executing river outfall remediation by implementing the "one mouth, one policy" program, achieving more than 70% of Taihu Lake Basin River outfall remediation tasks within the year, and initiating the long-term management of river outfalls (Yu et al. 2024). Undertaking comprehensive water pollutant remediation efforts for ships, advocating for the installation of domestic sewage storage facilities and transforming ship rubbish storage containers. Establishing a robust long-term regulatory mechanism for illicit inland waterway piers and enhancing the regional pipeline network and waste transfer service system for ports and piers (Yao et al. 2023). Intensifying pollution management for water-related enterprises. Leveraging special inspection actions for the emergency disposal facilities of water-related enterprises, making significant strides in promoting the remediation of corporate rainwater outfalls and emergency outfalls, and successfully rectifying issues with emergency disposal facilities for water-related enterprises by the end of June. Continuously advancing the standardization of phosphorus-related enterprises, engaging in standardized remediation, and incorporating phosphorus-related enterprises into the dynamic management list, with 50% of remediation tasks completed within the year.

12.4 Conclusion

- The built-up area sewage pipe network in Guxian Street has attained comprehensive coverage, ensuring the centralized treatment of regional sewage at the Garden Sewage Treatment Plant and effective control of regional water pollution discharge.
- The Guxian Street Sewage Treatment Project, particularly the Garden Sewage Plant Phase II, has reached a scale of 50,000 m^3/d. The current average daily capacity is 26,700 m^3/d, exceeding 80% of the first phase's designed scale of 30,000 m^3/d. Employing the "A/O contact oxidation + artificial wetland" treatment process, it follows a relatively centralized treatment mode. The tail water from the treatment facility meets the Class A standard outlined in the "Urban Sewage Treatment Plant Pollutant Discharge Standards" (GB18918-2002).

References

Chen M, Su Y, Piao Z, Zhu J, Yue X (2023) The green innovation effect of urban energy saving construction: a quasi-natural experiment from new energy demonstration city policy. J Clean Prod 428:139392

Li Y, Cao Y, Liu H, Li M, Xuan B, Zhang X, Gao X, Zhao J (2023) Contribution of wastewater treatment engineering measures in cities to reducing NH_3–N export to sea from subarea and river network perspectives using ecological network analysis: a case study of Fuzhou, China. Ocean Coast Manage 236:106501

Liu H, Zhang Z, Zhang T, Wang Z (2020) Revisiting China's provincial energy efficiency and its influencing factors. Energy 208:118361

Wang L, Duan X (2018) High-speed rail network development and winner and loser cities in megaregions: the case study of Yangtze River Delta, China. Cities 83:71–82

Xu H, Gao Q, Yuan B (2022) Analysis and identification of pollution sources of comprehensive river water quality: evidence from two river basins in China. Ecol Indic 135:108561

Xue W, Wang L, Yang Z, Xiong Z, Li XuQ, Cai Z (2023) Can clean heating effectively alleviate air pollution: an empirical study based on the plan for cleaner winter heating in northern China. Appl Energ 351:121923

Yao P, Wang Y, Liu J (2023) Can water pollution control influence employment adjustment in enterprises? Econ Anal Policy 80:384–397

Yu J, Xian Q, Cheng J, Chen J (2024) Horizontal ecological compensation policy and water pollution governance: evidence from cross-border cooperation in China. Environ Impact Asses 105:107367

Zhang Y, Huang G, Lu L (2015) Planning of water resources management and pollution control for Heshui River watershed, China: a full credibility-constrained programming approach. Sci Total Environ 524–525:280–289

Zhang H, Xing Y, Cheng S, Wang X, Guan P (2021) Characterization of multiple atmospheric pollutants during haze and non-haze episodes in Beijing, China: concentration, chemical components and transport flux variations. Atmos Environ 246:118129

Chapter 13
Justification of the Winter Temperature Regime of Water Hyacinth Content for Urban Wastewater Treatment

Vladimir Dmitrienko, Natalya Merenkova, Olga Pashkova, and Irina Zanina

Abstract Therefore, it determined an urgent need of reducing water pollution in urban wastewater facilities. The analysis of efficiency of domestic wastewater treatment in Shakhty of the Rostov region is given in the article. The negative impact of discharging treated domestic and groundwater into small rivers of the region is determined. The perspectives of additional treatment by means of higher aquatic plants (HAPs) are stated. The possibility of using water hyacinth as additional water purification is being considered. Taken into account the weather conditions which are unfavorable for Eichhornia, its vegetation period was described, which can reach up to 200 days. Based on the analysis of wastewater temperature measurements in municipal and mine wastewater treatment plants, favorable conditions for modeling of the water hyacinth in laboratory conditions were determined. While modeling conditions close to the water temperature in ponds of mine treatment plants, the conditions for survival of Eichhornia during the unfavorable winter time were determined. The most favorable temperature range for long-term storage of water hyacinth has been obtained allowing reducing the period of plants adaptation to natural conditions and active vegetation. Possible options of technical solutions for providing necessary conditions for preserving the thermophilic plant at low temperatures are determined. Among five wastewater treatment plants located in the town, the only one settling pond located at the Glubokaya mine can provide favorable temperature conditions for long-term storage of water hyacinth seedlings in winter. This will significantly increase the period of high-quality additional treatment of household wastewater.

Keywords Water pollution · Wastewater treatment · Water hyacinth · Temperature regime · Modeling

V. Dmitrienko (✉) · N. Merenkova · O. Pashkova · I. Zanina
Institute of Service and Business (branch) of DSTU in Shakhty, 346500 Shakhty, Rostov Region, Russia
e-mail: vadmitrienko@rambler.ru

Z. Wan (ed.), *Water Resources Management and Water Pollution Control*, Environmental Science and Engineering, https://doi.org/10.1007/978-3-031-53456-0_13

13.1 Introduction

Evolution on Earth would have been impossible without water, as the human body consists of 70% of water. Since ancient times, people have used water in all spheres of life, because clean water is necessary for human health. Almost three quarters of the Earth's surface is occupied by seas and oceans. Seawater accounts for 96.5% of the total volume, and the share of fresh water does not exceed 3.5%, but the volume available to humanity does not exceed 1% (Gerasimova 2016). Due to the water cycle in nature, a man used to consider water resources inexhaustible.

Urbanization of cities, rapid development of industry and agriculture in the twentieth century, land reclamation, and a number of other factors have led to problems with water supply for the population of the Earth due to the global nature of anthropogenic impact on the environment. Water consumption on the earth is 3,300–3,500 km^3 and 450 km^3 of domestic and industrial wastewater is discharged, which contains more than 13 thousand different elements and compounds (Lozanovska et al. 1998). At the same time, the number of natural sources of quality drinking water has been significantly decreasing over the last 30–40 years.

The increasing demand for water by industry and agriculture need make countries develop various measures for reducing water consumption and create new treatment technologies for preventing river pollution. Wastewater treatment is a complex technological task, as the equipment for currently known technologies is far from perfect and requires high construction costs. A considerable amount of money is required for personnel maintenance, equipment repair and so on.

13.2 Efficiency of Municipal Wastewater Treatment

Increasing of the efficiency of environmental protection can be ensured due to reducing pollution of air and water bodies. In other words, it is necessary to use waste-free technologies and transition to closed-loop water supply. However, in this case high quality of wastewater treatment is also required (Timofeeva 2001).

Currently, many methods and means of water treatment have been developed in the world. However, all of them have both advantages and disadvantages, i.e. it is either impossible or it is very expensive to purify water 100%. Therefore, the technological schemes of most existing wastewater treatment plants are multi-stage systems that allow reducing the proportion of pollutants in the treated water at each subsequent stage (Kulikova et al. 2020).

Even the first stage, which is relatively simple and designed for the removal of solid particles, consists in the sequential retention of suspended particles from large elements and fibrous compounds to micro filtering allows the removal of up to 90–95% of suspended solids from wastewater, while colloidal contaminants are only partially removed.

Semi-dissolved and dissolved substances, represented mainly by inorganic and organic chemical com-pounds require more complex and expensive equipment for additional purification. However, the analysis of treated domestic wastewater of our town shows that the degree of wastewater treatment is as follows: suspended solids—95.8%, ammonium ions—98.9%, phosphorus and phosphates—48.3%, nitrites—41.6%, dry residue—18.9% chlorides—10.7%, sulfates—14.8%. A more detailed study of the results of chemical analysis of the water at the outlet of the treatment facilities shows that when the air temperature decreases, the efficiency of treatment also decreases. That is, insufficiently treated water and its reuse in a closed cycle can be hardly possible even for technical purposes. In addition, there is a tendency of constant growth of nitrogen- and phosphorus-containing organic sub-stances in wastewater, so the degree of purification from nitrogen and phosphorus salts will decrease (Timofeeva 2001). Many re-searchers from around the world agree that domestic and industrial wastewater treatment facilities by means of using phytosystems shows good results (Artamonov 1986; Khramtsova et al. 1995; Dmitrieva and Einor 1995).

13.3 Prospects for Further Wastewater Treatment

The application of higher aquatic plants (HAPs) deserves special attention to perform deeper wastewater treatment (Stepanova and Simonova 2019; Kholodova and Rudikov 2019; Abuova et al. 2022). Many works are devoted to this theme. They contain the results of studies of many plants such as reeds, rushes, Urata, cassava, elodea, water sedge, aire and others. Such plants can accumulate and transform contaminants and contribute to additional water purification (Zhutov et al. 2010).

Despite the high efficiency of HAPs, they have serious disadvantages. All plants in our climatic conditions, when air and water temperature decreases, reduce their absorption capacity and even die. That is purification has a seasonal character. In addition, most of harmful substances can be accumulated in plants and require further utilization, which is not easy to perform in water environment. Therefore, it is possible to use floating plants, which occupy a special place in the bio purification systems, is the most appropriate decision.

The most important place in this group is occupied by the aquatic hyacinth (Eichhornia), which surpasses other plants in terms of vegetation speed and absorption of harmful impurities. However, it is not easy to keep thermophilic Eichhornia in winter, since this tropical plant is very sensitive to changes in climatic conditions.

13.4 Methods and Materials

Weather conditions in the central part of the Rostov region can be hardly suitable for the tropical plant, as the frosts can be registered in the second decade of September, and snow can be until mid-April for lately. That is, the unfavorable vegetation period of Eichhornia can reach 200 days.

The last decade is characterized by a significant increase in temperature in autumn, winter and spring. This means that there is a possibility to increase the duration of Eichhornia vegetation in open water bodies and, accordingly, to reduce costs for its preservation. In this case, it is necessary to study the possibility of providing minimum permissible conditions for preserving the viability of plants in the winter time.

Since the temperature regime of airflows depends on many factors in real conditions, it is extremely difficult to determine the influence of each of them on the vegetative ability of plants. Therefore, we decided to assess the viability of water hyacinth under prolonged exposure to low temperatures by simulating vegetation states in laboratory conditions.

To assess the temperature and humidity of the atmospheric air, an electronic thermometer TGC-MG4.01 was applied, which allows recording the measurement results in the device memory and then saving them on a computer. A special complex carried out continuous monitoring of air temperature with DS1820 temperature sensors, which transmitted data to the computer via1 Wire bus bar. The results of the long-term measurements were recorded in a database, which allowed the measurements to be plotted in graphs.

Operational control of water temperature directly at the treatment stages was carried out by a UT 71C meter, which also had the ability to save the results of observations.

13.5 Research Results

In order to identify the possibility to organize additional treatment of urban wastewater using water hyacinth, the technological scheme of urban wastewater treatment facilities, including three blocks: mechanical, biological treatment and sludge treatment, was considered. The temperature of wastewater from the wastewater disposal system of the city varies in the range of 17.6–19.4 °C, with an average flow rate of about 850 m^3/hour.

The above-mentioned temperature allows cultivation of Eichhornia all year round. However, the difficulty lies in the fact that the concentration of pollutants up to the outlet of the effluent from the aero tank of the biological treatment unit significantly exceeds the maximum permissible for water hyacinth, despite the fact that it can process them in large quantities. In addition, in January–February the water temperature decreases to +15 °C as it passes through a number of facilities during treatment.

The water then enters secondary radial settling tanks with a diameter of 28 m each. Only in these sedimentation tanks the concentration of pollutants reaches the values permissible for Eichhornia vegetation. However, the technology of operation of these structures does not allow cultivation of plants in them. In addition, at the outlet of radial settling tanks the water temperature in January can drop to 11 °C and below.

Thus, the results of the research showed that the temperature of treated water does not fall below +10 °C. The air temperature is much more complicated. If up to mid-October the average daily temperature is within 11–15 °C (Fig. 13.1), then already at the end of October relatively long periods with a decrease to negative temperatures are observed (Fig. 13.2).

In November–December, the average temperature is characterized by frequent transitions to zero, but on some days even in November it can drop below −12 °C. Stable negative temperatures persist in January and February, but sometimes in February, it is possible to observe an increase to +13 °C. The period of March–April is also characterized by significant temperature variations during the day and at night with frequent frosts. Only by the first decade of May, a stable increase to 13–20 °C can be observed (Fig. 13.3).

Thus, observations during 2018–2023 show that despite the warming trend, the duration of an unfavorable period for the vegetation of an aquatic hyacinth can be 180–200 days. Therefore, the preservation of plants for wastewater treatment during this period is a difficult and expensive task.

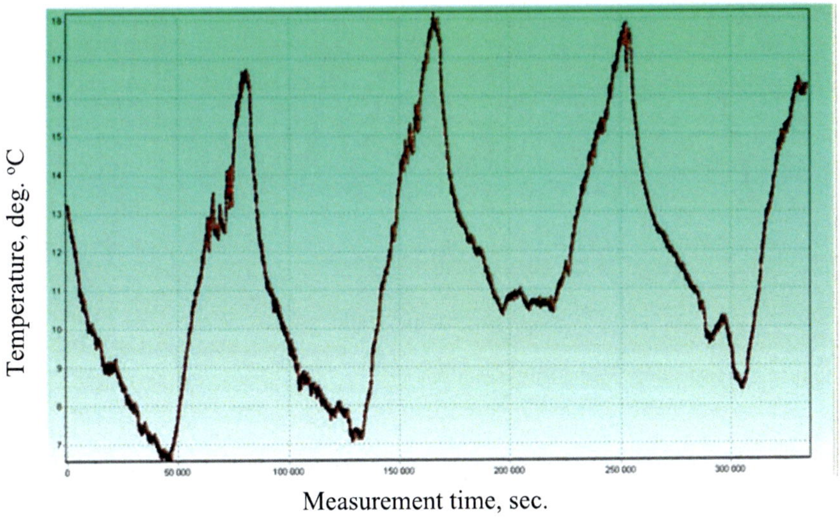

Fig. 13.1 Thermo gram for 8–12 October

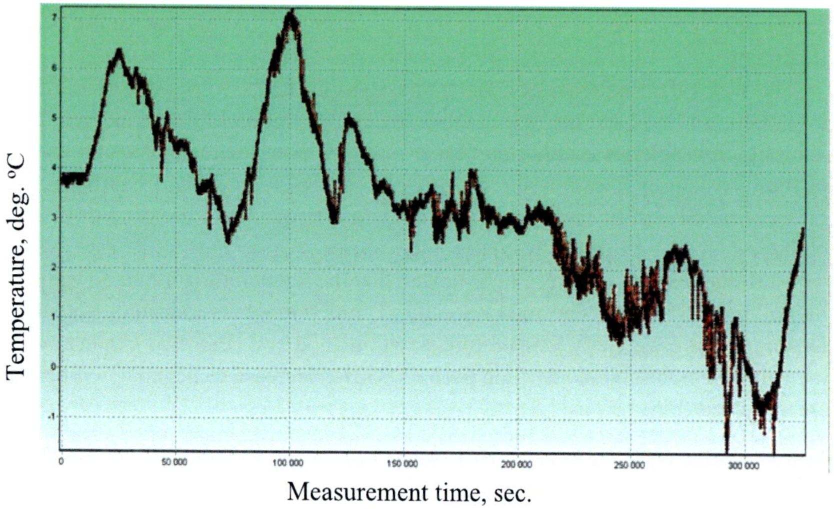

Fig. 13.2 Thermo gram for 21–25 October

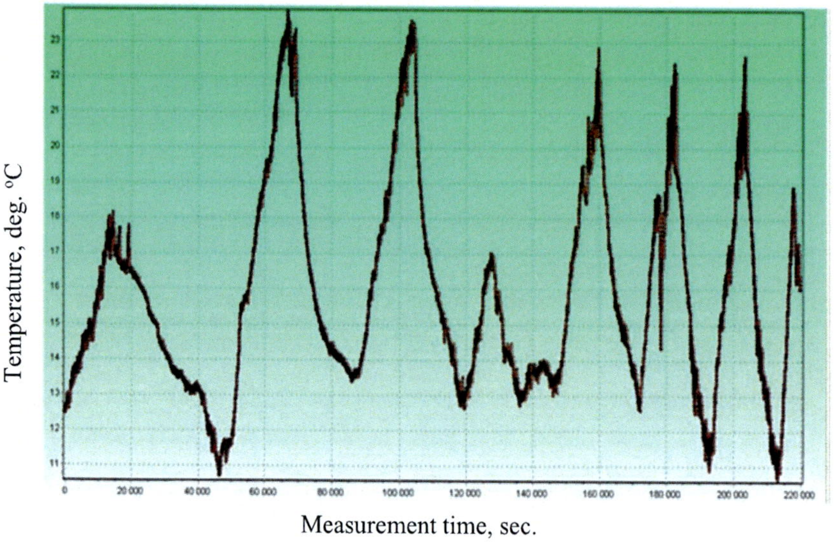

Fig. 13.3 Temperature fluctuations in May 2021

A promising option for preservation of water hyacinth in cold periods of the year may be its preservation in the ponds of mine water settling ponds, the temperature of which even in the coldest days is: at the mine "Glubokaya" is 18 °C, "Yuzhnaya" is 12–14 °C, "Mayskaya" is from 8 to 10 °C.

Fig. 13.4 Placement of water hyacinth in the laboratory

This choice is based on the fact that hundreds of cubic meters of water are pumped from great depths, so the temperature of the water in the aerators remains constant throughout the year (Kulikova et al. 2020). The large area and volume of settling ponds ensure laminar flow and very high thermal inertia not only of water but also of air near the surface.

In order to determine the viability of water hyacinth at such temperatures, four batches 1, 2, 3, 4 of approximately the same weight and appearance of plants were formed in the laboratory. Each batch was placed in two plastic containers with a surface area of 0.13 m^2 (Fig. 13.4).The water level was maintained within 12–15 cm.

Plants were placed in the laboratory on 22.10.2022. Storage was carried out in specially made cabinets at temperatures of (1) 24–26, (2) 17–18, (3) 12–13 and (4) 8–10 °C.

For all batches of plants, the same combined lighting was used, consisting of 2 × 30 W fluorescent lamps and 18 W full-spectrum LED lamps for plants, located at a distance of 20–35 cm from the leaf surface. The duration of illumination was regulated by a timer and was 11 h. Humidity was maintained at 55–70%.

13.6 Discussion of the Results

In March 2023, a check showed that the plants stored at 24–26 °C looked best, as new shoots and even flowers were present (Fig. 13.5). At 17–18 °C, the plants were stunted but looked viable. For the batch stored at 12–13 °C, it was necessary to place the plants on floats, as they began to sink and die. The batch stored at 8–10 °C was almost dead by the beginning of March (Fig. 13.6).

On 11.04 2023, the plants were placed in open pools. However, until the beginning of May, the pools were periodically covered with non-woven material at night due to temperature decrease.

Fig. 13.5 The state of the water hyacinth at the temperature of 24–26 °C

Fig. 13.6 The state of water hyacinth at the temperature of 8–10 °C

Plants with the highest storage temperature adapted in a short time and grew rapidly. The batch with 17–18 °C required a longer adaptation period, but then the vegetation rate was quite high (Fig. 13.7).

The temperature of 12–13 °C has a detrimental effect on water hyacinth during long-term storage and the adaptation period can reach 40 days. New shoots develop very slowly, although flowering is observed (Fig. 13.8).

Thus, the possibility of long-term storage of water hyacinth in winter at temperatures of mine water drainage is experimentally confirmed, if protection from external airflows is provided.

Fig. 13.7 Vegetation of water hyacinth in July 2023, stored in winter at 17–18 °C

Fig. 13.8 The vegetation of water hyacinth in July 2023, stored in winter at 12–13 °C

13.7 Conclusions

For normal vegetation of water hyacinth, a certain temperature regime, lighting and chemical composition of water are required. The performed studies show that in the absence of nutrition, in conditions of artificial lighting, but with a duration of at least 11 h, water hyacinth can remain viable in the pond of treatment facilities of the mine "Glubokaya", because the water temperature in it does not fall below 18 °C.

The air temperature at the water surface may change depending on the speed of wind, the intensity of solar radiation and the place of measurements in the pond, but negative values were not noted.

Water bodies of municipal treatment facilities and groundwater at the "Mayskaya" and "Yuzhnaya" mines are not suitable for water hyacinth storage without additional measures.

Chemical analysis of mine waters shows a significant excess of the content of a number of substances, in particular iron. At the same time, basic nutrients such as nitrogen, phosphorus, potassium and sodium are practically absent in mine water, therefore, it is advisable to place water hyacinth in sealed floating containers with protection from air currents. This will make it possible to increase the vegetation period of the plant by planting "adult" plants in spring at additional treatment facilities at low financial cost.

Plants and seeds that died after crystallization of water did not germinate throughout the summer period, that is, the possibility of water hyacinth clogging the rivers into which purified water is discharged is unlikely.

References

Abuova GB, Kharlamova AE, Sardina A, Sardina S (2022) Efficiency of water hyacinth (eichornia crassipes) application in wastewater pretreatment. Engineering and Construction Bulletin of the Caspian Sea: scientific and technical journal. Astrakhan: SAEU AO VO "AGASU" 1(39):33–37

Artamonov VI (1986) Plants and purity of the natural environment. Science, Moscow, p 172

Dmitrieva NG, Einor LO (1995) The role of macrophytes in the transformation of phosphorus in water. Water Resour Manage 5:101–110

Gerasimova AE (2016) Problem of fresh water depletion. Youth and scientific and technological progress. In: IX International scientific and practical conference of students, postgraduates and young scientists: in 4 volumes, pp 207–209

Kholodova SN, Rudikov DA (2019) On the possibility of using water hyacinth for purification of polluted water. Water Ecol Probl Solutions 3(79):70–76

Khramtsova TG, Stom DI, Vigoda VA (1995) The use of macrophytes for post-treatment of municipal wastewater. Probl Ecol 2:260–262

Kulikova AA, Sergeeva YA, Ovchinnikova TI, Khabarova EI (2020) Formation of mine waters and analysis of methods of their treatment. Min Inf Anal Bull 7:135–145

Lozanovska IN, Orlov DS, Sadovnikova LK (1998) Ecology and protection of the biosphere in chemical pollution: studies benefit. Higher School, Moscow, p 287

Stepanova SA, Simonova GV (2019) Water hyacinth—a natural water purifier. Vestnik SGUGiT 24(1):264–267

Timofeeva SS (2001) Biotechnology of wastewater treatment. Water Chem Tech 5:525–532

Zhutov AS, Rogacheva SM, Gubina TI (2010) Investigation of the possibility of desalination of the Balakovo NPP cooling pond with the help of higher aquatic plants. In: Izvestiya Samara Scientific Centre of the Russian Academy of Sciences, no 1(8), vol 12, pp 2125–2128

Chapter 14
Numerical Simulation of Ba^{2+} Transport in Vertical Plastic Concrete Cutoff Barrier

Jintao Yao, Haoqing Xu, Jingrui Liang, Wenyang Zhang, and Shaowen Hou

Abstract Taking the vertical plastic concrete cutoff barrier in the Guizhou Tianzhu Chemical Phase II landfill as the research object, the Hydrus-2D finite element simulation software was used to investigate the migration law of barium ion. Firstly, a plastic concrete soil column model was established, and the initial conditions of barium ion concentration, hydraulic head, and the boundary conditions of barium ion concentration were set. The software simulated the temporal and spatial variations of barium ion concentration in the vertical plastic concrete cutoff barrier, and evaluated the barrier effect of the vertical plastic concrete cutoff barrier on barium ions. The study revealed the following: (1) The vertical plastic concrete cutoff barrier exhibits a good barrier effect on the migration of barium ions. (2) The fastest migration of barium ions in the vertical plastic concrete cutoff barrier occurs at a depth of 22 m from top to bottom, making it more prone to breakthrough by barium ions. (3) For safety reasons, a barium ion concentration of 4 mg/L is recommended as the breakthrough criterion in the engineering project.

Keywords Plastic concrete · Cutoff barrier · Barium ion · Pollutant transport

J. Yao (✉) · H. Xu (✉) · J. Liang · W. Zhang · S. Hou
College of Civil Engineering and Architecture, Jiangsu University of Science and Technology, Zhenjiang 212028, China
e-mail: 212210901133@stu.just.edu.cn

H. Xu
e-mail: hankinxu@163.com

H. Xu
Jiangsu Geological Environment Disaster Prevention and Restoration Engineering Research Center, Zhenjiang 212028, China

Z. Wan (ed.), *Water Resources Management and Water Pollution Control*, Environmental Science and Engineering, https://doi.org/10.1007/978-3-031-53456-0_14

14.1 Introduction

With the rapid development of China's economy, barite as an important mineral resources, its exploitation degree is increasing. China is rich in barite resources, especially in mountainous areas (Wang et al. 2023; Liu et al. 2021). However, in the early production process, some enterprises ignored the protection of the environment, resulting in the waste generated in the production process is directly stacked in the temporary storage yard. Due to the geological characteristics of mountainous areas, cracks are developed and rain is abundant. Heavy metal ions in these waste residues will gradually dissolve with the action of rain and enter the nearby water bodies (Wang et al. 2016). This has led to a series of problems such as environmental pollution and harm to human health (Fan 2020; Chen et al. 2006). Heavy metal pollution in waste residues is a problem that cannot be ignored, and it has potential impacts on the ecosystem and the health of residents.

In order to ensure people's safety, the vertical plastic concrete cutoff barrier is generally set around the storage yard to prevent seepage and intercept pollution (Zhang 2005). The vertical plastic concrete cutoff barrier has been studied by many experts.

Wu, J.Z. (2018) studied the impermeability of plastic concrete from three aspects: water-binder ratio, cement dosage and water-reducing agent dosage respectively through the relative permeability coefficient method, and found that with the increase of cement dosage and the decrease of water quantity, the permeability coefficient of plastic concrete would decrease significantly. Wang, G.W. (2023) studied the early compressive strength (3 and 7 days) and splitting tensile strength of concrete with different metakaolin content, and used scanning electron microscopy (SEM) to study the morphology of hydration products of concrete with different metakaolin content at 3 and 7 days of curing age. The early strength results show that with the increase of metakaolin content, the compressive strength and splitting tensile strength of the prepared plastic concrete in the early stage increase at first and then decrease, and the optimal metakaolin content is 10% of the total cement mass. The SEM structure shows that the proper amount of metakaolin can increase the compactness between the early hydration products, which leads to the increase of the early strength of plastic concrete. Li, G.B. (2023) discussed the influence of 10%, 20%, 30% and 40% ore powder content on the performance of plastic concrete by laboratory tests, and further analyzed the action mechanism of ore powder. The results show that the strength of plastic concrete is the highest when it is mixed with 30% silica powder, and the strength of concrete can be increased in different degrees by adding appropriate amount of silica powder. Under the same dosage, the effect of silica powder on improving the splitting tensile strength is the most obvious, followed by the tensile strength, and the effect on improving the bending strength is the weakest. The elastic modulus and strength of plastic concrete increase first and then decrease with the increase of silica powder content, and the increase of compressive strength is higher than the elastic modulus, which can effectively reduce the elastic-strength ratio and relative permeability coefficient. Wu, J.F. (2022) explored the effects of coarse aggregate content on compressive strength, water permeability and failure mode of plastic concrete through laboratory tests, and obtained the optimal mix ratio. The results show that the strength of concrete increases obviously when the content of coarse aggregate is 770.8 ~ 1028.7 kg/m^3. When the coarse aggregate content is 1120.6 ~ 1230.8 kg/m^3, the concrete strength deteriorates. With the increase of coarse aggregate content, the permeability of concrete increases gradually.

Yu, F.P. (2023) explained the construction technology of plastic concrete cut-off wall through the rescue and reinforcement project of Yanqian reservoir. The use of plastic concrete cure-off wall construction technology can effectively realize the reinforcement treatment of reservoir projects, because the concrete cure-off wall has a strong anti-seepage capacity, and its low elastic modulus can effectively reduce the internal stress of the cure-off wall, so as to effectively improve the quality and stability of the cure-off wall, and further extend the service life of the dam body of the reservoir. Based on the example of seepage prevention and reinforcement of earth dam of Da qin Reservoir in Guangdong province, Lai, Z.R. (2023) proposed six control factors for the whole construction process of plastic concrete wall from the aspects of pilot hole, concrete mix ratio, tank quality, tank cleaning and slurry changing quality, concrete pouring quality and joint section treatment quality. During the construction of the cutoff wall, the whole process should be controlled, starting from the early preparation to the end of the entire project construction, and the process should be strictly controlled. Especially for the troughing and pouring process of plastic concrete wall, it is necessary to strengthen the control. Zheng, L.S. (2022), Chang, D.M., et al. (2022) discussed the construction technology of plastic concrete cutoff wall based on the actual project, and compared and selected the schemes, providing reference for the construction of related projects.

In current engineering practice, the vertical plastic concrete cutoff barrier is widely used to prevent groundwater and soil from being damaged by pollutants. However, although its anti-seepage performance has been widely studied and applied (Qian et al. 2017), relatively little research has been done on its pollution interception and pollutant transport characteristics. This is a pressing issue because we need a more comprehensive understanding of how to most effectively capture and isolate subsurface contaminants, especially important contaminants such as barium ions.

Due to the relatively long diffusion and migration time of Ba^{2+} in the vertical plastic concrete cutoff barrier, it is difficult to completely track and analyze the migration of Ba^{2+} in the limited laboratory conditions. The finite element software can make up for the shortcomings of traditional laboratory tests. It adjusts the simulation time to simulate this process according to the parameters obtained from the test and the actual boundary conditions, and its analysis results are more valuable for reference (Xu et al. 2021). Commonly used numerical analysis software include GeoStudio, Hydrus-2D, GMS, Visual Modflow and so on. Lv, J.J., Shi, L.J., et al. (2022) applied Hydrus-1D software to simulate the process of carbon, nitrogen and phosphorus removal in soil percolation system improved by micro-nano aeration. Dong, S.J., CAI, T.Z., et al. (2023) used Hydrus-1D software to build a soil unsaturated zone water flow model and solute transport model to simulate and predict the migration depth, migration time and concentration changes of petroleum hydrocarbon pollutants in soil environment. Roger, Liu, B.H., et al. (Luo et al. 2023) applied the Hydrus-1D software to establish a heavy metal solute transport model to simulate the spatiotemporal change process of distribution and transport of heavy metal solute in soil, and obtain its maximum infiltration distance in different time periods and concentration equilibrium time in different soil layers. Fu, Z., Zhang, F.C., et al. (2021) used the Hydrus-1D model to simulate the characteristics of soil heavy metal pollution, and proposed remediation measures based on the simulation results. Practice results show that the results are accurate, effectively shorten the pollution treatment period, avoid the damage of the traditional multi-point drilling sampling method to the reservoir water barrier layer, reduce the risk of leakage, and help to ensure the safety of water quality.

In this paper, Hydrus-2D software is used to establish a two-dimensional seepage Ba^{2+} migration model, and the barrier effect of plastic concrete vertical anti-fouling barrier is analyzed. Based on the parameters of each soil layer in the geological survey report, the geometric size, permeability coefficient, dispersion coefficient, diffusion coefficient, adsorption allocation coefficient and other calculation parameters of each soil layer and barrier were determined, and a two-dimensional pollutant migration model was established to simulate the relationship between pollutant concentration and time and space in the presence of vertical plastic concrete cutoff barrier. To evaluate the effect of vertical plastic concrete cutoff barrier on pollutants, and to provide a reliable theoretical basis for practical engineering application.

14.2 Hydrus-2D Software Simulation Theory

The Hydrus software is designed to simulate the movement of water, heat and solutes in saturated and unsaturated porous media. In porous media, the Richards equation of saturated–unsaturated flow and the convection–dispersion equation are used to control the transport of heat and solute. For solute transport, the software supports constant concentration and concentration flux boundaries. The software provides Van Genuchten model, Corey model and modified Van Genuchten model to describe the hydraulic characteristics of unsaturated soil (Simunek et al. 2006).

14.2.1 *Soil Water Movement Theory*

Soil water movement is mainly controlled by Richards equation (Ministry of Ecological Environment 2019).

$$\frac{\partial \theta}{\partial t} = \frac{\partial}{\partial x}\left[D\frac{\partial \theta}{\partial x}\right] + \frac{\partial}{\partial z}\left[D\frac{\partial \theta}{\partial z}\right] + \frac{\partial K}{\partial z} \tag{14.1}$$

where θ is the water content, D is the diffusion coefficient, and K is the unsaturated permeability coefficient.

The θ and K of soil hydrodynamic properties in Eq. (14.1) are modified by Van Genuchten equation (Van Genuchten 1980). This equation can describe the hydrodynamic properties of soil more flexibly when the soil water is close to saturation. The equation is as follows:

$$\theta = \begin{cases} \theta_r + \dfrac{\theta_s - \theta_r}{\left[1 + |\alpha h|^n\right]^{1-1/n}}, h < 0 \\ \theta_s, h \geq 0 \end{cases} \tag{14.2}$$

$$S_e = \frac{\theta - \theta_r}{\theta_s - \theta_r} \tag{14.3}$$

$$K = K_s S_e^l \left[1 - \left(1 - S_e^{n/n-1}\right)^{1-1/n}\right]^2 \tag{14.4}$$

where, S_e is the effective saturation of soil; θ_s is the saturated water content of soil. θ_r is the residual water content of soil. K_s is the saturation permeability coefficient; α, n and l are empirical parameters.

14.2.2 Control Equation of Solute Transport

The increase of Ba^{2+} mass comes from four aspects, one is through dispersion, the other is through convection, the third is through source-sink and the fourth is adsorption (Wang 2008). The variation of Ba^{2+} mass in equilibrium unit is caused by dispersion, convection, source-sink and adsorption. According to the law of conservation of mass,

$$\Delta M' = \Delta M + \Delta N + \Delta M_l + \Delta M_a \tag{14.5}$$

Ignore the deformation of the medium and get,

$$\begin{aligned} \frac{\partial nC}{\partial t} &= \frac{\partial}{\partial x}\left(nD_{xx}\frac{\partial C}{\partial x} + nD_{xy}\frac{\partial C}{\partial y} + nD_{xz}\frac{\partial C}{\partial z}\right) \\ &+ \frac{\partial}{\partial y}\left(nD_{yx}\frac{\partial C}{\partial x} + nD_{yy}\frac{\partial C}{\partial y} + nD_{yz}\frac{\partial C}{\partial z}\right) + \frac{\partial}{\partial z}\left(nD_{zx}\frac{\partial C}{\partial x} + nD_{zy}\frac{\partial C}{\partial y} + nD_{zz}\frac{\partial C}{\partial z}\right) \\ &- \frac{\partial nu_x C}{\partial x} - \frac{\partial nu_y C}{\partial y} - \frac{\partial nu_z C}{\partial z} \\ &+ I - \rho_b \frac{\partial C_s}{\partial t} \end{aligned} \tag{14.6}$$

This is the Ba^{2+} transfer differential equation. Where C is the concentration of Ba^{2+}; D_{xx} D_{xy} D_{xz} D_{yx} D_{yy} D_{yz} D_{zx} D_{zy} D_{zz} is the coordinate component of the hydrodynamic dispersion tensor; $u_x u_y u_z$ is the actual velocity of groundwater movement; I is the source and sink item; n is the porosity.

According to the definition of blocking factor,

$$R_d = 1 + \frac{\rho_d \partial C_s}{n \partial C} \tag{14.7}$$

Change the above formula to,

$$R_d \frac{\partial nC}{\partial t} = \mathrm{div}(nD\mathrm{grad}C) - \mathrm{div}(nuC) \tag{14.8}$$

where div is the divergence operator; grad is the gradient operator.

14.3 Two-Dimensional Ba^{2+} Migration Numerical Simulation

14.3.1 Establishment of Finite Element Calculation Model

In order to meet the design requirements, the vertical anti-seepage system and the waterproof curtain of the vertical plastic concrete cutoff barrier should be embedded down the impervious layer. According to the survey data, the permeability coefficient k of the sixth layer of breezy formation is $1.6 \times 10^{-7} \sim 9.5 \times 10^{-6}$ cm/s, and that of the seventh layer of unweathered rock is $8.6 \times 10^{-9} \sim 9.5 \times 10^{-8}$ cm/s, and the formation distribution is relatively uniform. Therefore, the 7th layer of unweathered rock is used as the lower impervious layer of the vertical plastic concrete cutoff barrier, and the 6th layer of breezy formation is used as the lower relative impervious layer of the water seal curtain, and the embedding depth is 2 m. With reference to the survey data, a two-dimensional model of pollutant migration at the site was established (see Fig. 14.1).

14.3.2 Calculation Parameters

14.3.2.1 Soil–Water Characteristic Parameters

The experimental data show that the empirical parameters of soil–water characteristic curve of plastic concrete range from $2.2 \sim 5.4 \times 10^{-5}$, from 1.31 to 2.05, from 0.012 to 0.096, from 0.2966 to 0.3840. Therefore, the empirical parameters, empirical parameters n, residual volume water content and saturated volume water content of

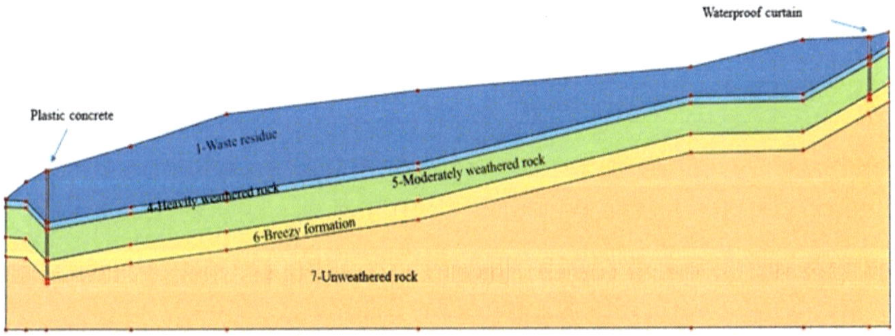

Fig. 14.1 Finite element calculation model

Table 14.1 Soil and water characteristic parameters of each soil layer

Level number	Soil layer name	α	n	θ_s	θ_r
1	Waste residue	0.08	1.40	0.50	0.07
4	Heavily weathered rock	0.06	1.70	0.35	0.05
5	Moderately weathered rock	0.05	1.75	0.30	0.05
6	Breezy formation	0.04	1.80	0.25	0.04
7	Unweathered rock	0.03	1.85	0.20	0.02
	Plastic concrete	0.00004	1.70	0.35	0.05
	Waterproof curtain	0.05	1.60	0.35	0.06

the soil–water characteristic curve of plastic concrete are set as 4×10^{-5}, 1.7, 0.05 and 0.35 respectively.

Waste residue is considered as weak soil in the geological exploration report, but in order to simplify and be more reasonable, this paper equates waste residue as soil mass and obtains the soil and water characteristic parameters of waste residue (Carsel and Parrish 1988). With reference to the data of soil–water characteristic test of weathered granite (Peranić et al. 2018), the soil–water characteristic parameters of rock layers with different weathering degrees are set, as shown in Table 14.1.

14.3.2.2 Permeability Coefficient

The experimental data show that the permeability coefficient of plastic concrete enters the stable stage after 28 days, and the permeability coefficient of plastic concrete is between 1.8 ~ 6.8 × 10^{-8} cm/s. Therefore, the permeability coefficient of plastic concrete is set to 4.00 × 10^{-8} cm/s. Water pressure test was carried out on each soil layer to obtain the permeability coefficient of each soil layer, as shown in Table 14.2.

Table 14.2 Permeability coefficient of each formation

Level number	Soil layer name	Permeability coefficient (cm/s)
1	Waste residue	5.10×10^{-3}
4	Heavily weathered rock	2.00×10^{-4}
5	Moderately weathered rock	6.10×10^{-5}
6	Breezy formation	4.40×10^{-6}
7	Unweathered rock	6.71×10^{-8}
	Plastic concrete	4.00×10^{-8}
	Waterproof curtain	2.00×10^{-7}

14.3.2.3 Dispersion, Diffusion Coefficient and Adsorption Allocation Coefficient

The experimental data show that the effective diffusion coefficient of Ba^{2+} in the soil sample in this study is 1 ~ 2.89×10^{-10} m^2/s, so the effective diffusion coefficient of Ba^{2+} in plastic concrete is set at 2×10^{-10} m^2/s. The longitudinal dispersion is between 0.001 and 0.01. Considering that plastic concrete will have defects in the actual construction process and the project is located outdoors, the longitudinal dispersion is set at 0.1 and the transverse dispersion is set at 0.01 (Van Genuchten 1980). The adsorption and distribution coefficient of Ba^{2+} obtained in the convection–diffusion test ranges from 6.316 to 6.920. Considering the influence of adsorption and distribution coefficient, the adsorption and distribution coefficient is set as 6.5 here.

The main parameters of the soil layer are shown in Table 14.3. The longitudinal and transverse dispersion of each soil layer are set according to the empirical data (Wang 2008), the diffusion coefficient of the soil layer refers to the statistical data (Xie 2008), and the adsorption allocation coefficient is set according to the empirical value (Qian et al. 2017).

14.3.2.4 Initial Conditions

1. Initial conditions for concentration

The concentration of Ba^{2+} in the site leachate ranged from 1232 mg/L to 4969 mg/L. Considering the uneven distribution of pollutants, the initial concentration of pollutants in this area was set as 5000 mg/L. In addition, the water storage capacity of the rock is poor, so it is assumed that the pollutants are all distributed in the waste slag, and there are no pollutants in the rock formation.

2. Initial conditions of water head

Analyze the topographic conditions provided by the geological survey report and select suitable water levels upstream and downstream. The downstream water level is 2.5 m underground, and the upstream water level is 25 m underground.

Table 14.3 Main parameters of each soil layer

Level number	Soil layer name	Longitudinal dispersion	Lateral dispersion	Diffusion coefficient (m^2/s)	K_d (L/kg)
1	Waste residue	0.1	0.01	6.00×10^{-10}	1.0
4	Heavily weathered rock	0.1	0.01	3.00×10^{-10}	2.4
5	Moderately weathered rock	0.1	0.01	2.50×10^{-10}	2.6
6	Breezy formation	0.1	0.01	1.80×10^{-10}	2.8
7	Unweathered rock	0.1	0.01	1.50×10^{-10}	3.0
	Plastic concrete	0.1	0.01	2.00×10^{-10}	6.5
	Waterproof curtain	0.1	0.01	2.00×10^{-10}	2.5

14.3.2.5 Boundary Conditions

In a pollutant site, since the concentration of pollutants in groundwater is constant, the first type of concentration boundary conditions is set at the upstream boundary:

$$c(x, y, t)|_{\Gamma_1} = c_1(x, y, t), (x, y) \in \Gamma_1 \tag{14.9}$$

Since the downstream boundary is on the other side of the cutoff wall, there are no pollutants in the formation, so the downstream concentration boundary condition is set to a constant concentration boundary:

$$c(x, y, t)|_{\Gamma_2} = 0, (x, y) \in \Gamma_2 \tag{14.10}$$

14.4 Analysis of Numerical Simulation Results

14.4.1 Seepage Calculation Results and Analysis

Figure 14.2 from day 0 to day 3,700, the water flow passed through the bottom of the water stop curtain and continued to flow downstream, but did not reach the vertical plastic concrete cutoff barrier. From the upper reaches to the lower reaches, the water flow speed increases and then decreases, and then increases and then decreases. On both sides of the water stop curtain, the seepage velocity changes sharply, which

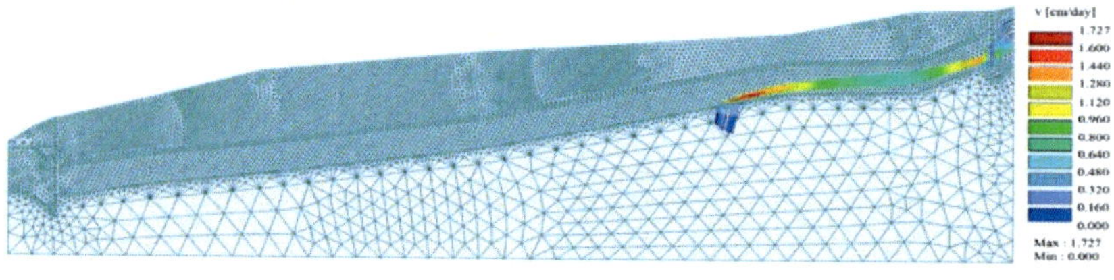

(a) Seepage on day 3700

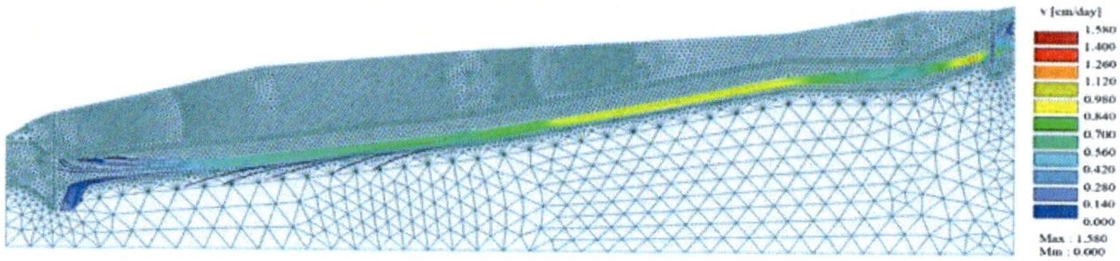

(b) Seepage on day 11000

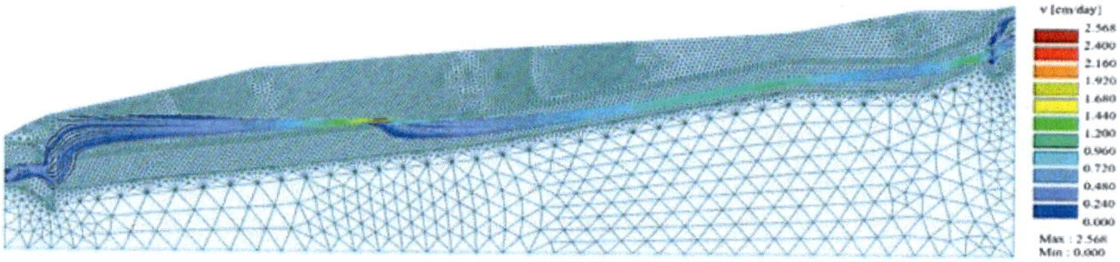

(c) Seepage on day 36500

Fig. 14.2 Seepage calculation results. **a** Seepage on day 3700, **b** Seepage on day 11,000, **c** Seepage on day 36,500

indicates that the water stop curtain has a certain blocking effect on the seepage, so it is reasonable to arrange the water stop curtain above the site

From day 3,700 to day 11,000, the water flow, although blocked by the water curtain above the site, was able to continue to flow down the site and reach the vertical plastic concrete cutoff barrier. In addition, although the water reaches the vertical plastic concrete cutoff barrier, it does not pass through the barrier. Because the permeability coefficient and diffusion coefficient of the plastic concrete cuttable wall are small, the permeability of the barrier is poor, and the permeability resistance of the barrier is strong, the water flow cannot pass through the plastic concrete vertical anti-fouling barrier in a short time. This shows that the vertical plastic concrete cutoff barrier is effective for water flow.

From day 11,000 to day 36,500, the water gradually passes through the lower part of the vertical plastic concrete cutoff barrier at a very low flow rate and penetrates into the vertical plastic concrete cutoff barrier. Because the hydraulic gradient in the middle and upper part of the vertical plastic concrete cutoff barrier is relatively small, the permeability coefficient is relatively small, the diffusion coefficient is relatively small, and the permeability is small, so the seepage above the barrier cannot pass through. Under the vertical plastic concrete cutoff barrier, the hydraulic gradient is relatively large, the permeability coefficient is relatively large, the diffusion coefficient is relatively large, and the permeability is large, so the water flow will pass through the lower part of the vertical plastic concrete cutoff barrier at a lower speed, and the flow around the bottom of the anti-pollution barrier will be generated. With the passage of time, the seepage continues to expand and reach a stable state in the middle and lower part of the vertical plastic concrete cutoff barrier.

14.4.2 Calculation Results and Analysis of Ba^{2+} Migration

14.4.2.1 Overall Migration Situation

Figure 14.3 from day 0 to day 3,700, Ba^{2+}, driven by water flow, continuously migrated to the weathered rock strata and cutoff barrier. Under the action of water flow, Ba^{2+} inside the water curtain above the site continuously migrates to the downstream of the site, resulting in the decrease of Ba^{2+} concentration inside the water curtain. Because the rock layer is subjected to weathering for a long time, a small amount of Ba^{2+} enters the weathered rock layer under the waste residue. For the downstream plastic concrete vertical anti-fouling barrier, Ba^{2+} has almost no migration in the vertical plastic concrete cutoff barrier due to the small hydraulic gradient and the small permeability coefficient and diffusion coefficient of the vertical plastic concrete cutoff barrier, which has a blocking effect.

From day 3700 to day 11,000, Ba^{2+} continues to migrate towards the vertical plastic concrete cutoff barrier under the action of water flow. Since the water has reached the cutoff barrier at this time, the upstream part of Ba^{2+} is carried downstream, resulting in a slight increase in the concentration of Ba^{2+} inside the cutoff barrier. However, the permeability coefficient and diffusion coefficient of the vertical plastic concrete cutoff barrier are very small and have retarding effect, so Ba^{2+} has almost no migration in the vertical plastic concrete cutoff barrier. Due to the long-term weathering of underground rock layers, trace amounts of Ba^{2+} will enter the moderately weathered strata over time.

Between day 11,000 and day 36,500, Ba^{2+} gradually reached the downstream of the site and the plastic concrete emerged from the cutoff barrier. Due to seepage, Ba^{2+} in the waste slag in the middle and lower part of the site will migrate with water flow, resulting in a decrease in local Ba^{2+} concentration in the lower part of the lower part and an increase in local Ba^{2+} concentration, which reaches the maximum in the inner part of the vertical plastic concrete cutoff barrier. The vertical plastic concrete cutoff barrier blocks the outward migration of Ba^{2+}, which results in the accumulation of Ba^{2+} in the interior of the vertical plastic concrete cutoff barrier. Only a small amount of Ba^{2+} gradually migrates through the vertical plastic concrete cutoff barrier under the action of water flow.

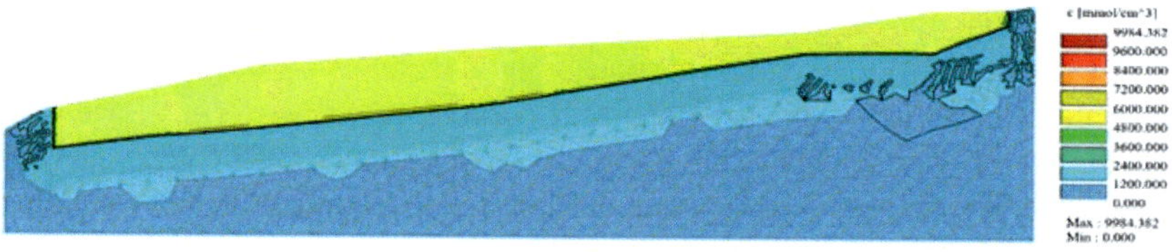

(a) Chart of Ba^{2+} migration on day 3650

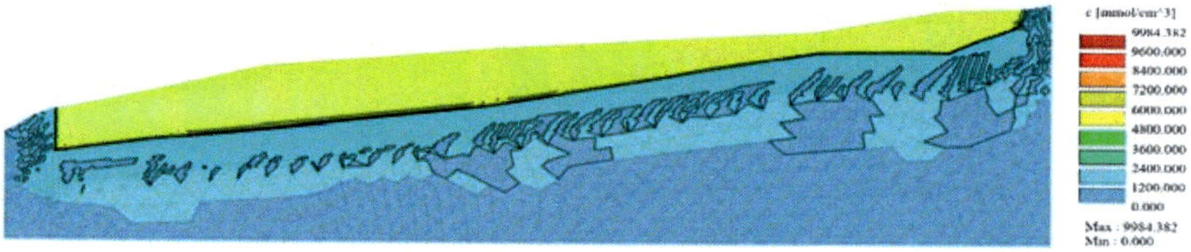

(b) Chart of Ba^{2+} migration on day 11000

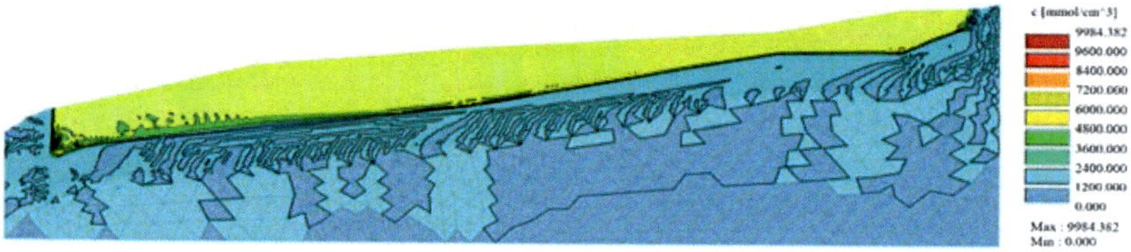

(c) Chart of Ba^{2+} migration on day 36500

Fig. 14.3 Barium ion migration diagram. **a** Chart of Ba^{2+} migration on day 3,650, **b** Chart of Ba^{2+} migration on day 11,000, **c** Chart of Ba^{2+} migration on day 36,500

14.4.2.2 Migration of Observation Points

In order to further evaluate the cutoff effect of the vertical plastic concrete cutoff barrier, 19 observation points are set on the outer side of the vertical plastic concrete cutoff barrier from the top down, with a distance of 2 m. The use effect of the cutoff barrier is evaluated by the migration of pollutants at the observation point at different times.

Through finite element calculation, the observation data of 19 observation points were obtained. Since the Ba^{2+} concentration from 0 to 8 m and from 30 to 36 m was basically 0 mg/L, only the observation data from 10 to 28 m were given here (Fig. 14.4).

Figure 14.5(a) from 10 to 22 m, Ba^{2+} concentration at the observation point increases with increasing depth, reaching the maximum value at 22 m; Fig. 14.5(b) From 22 to 28 m, the Ba^{2+} concentration at the observation point becomes smaller and smaller as the depth becomes deeper and deeper, reaching the maximum value at 22 m. It can be concluded that Ba^{2+} migrates the fastest in the vertical plastic concrete cutoff barrier at 22 m, which is the weakest part of the cutoff barrier and the most easily broken down by Ba^{2+}.

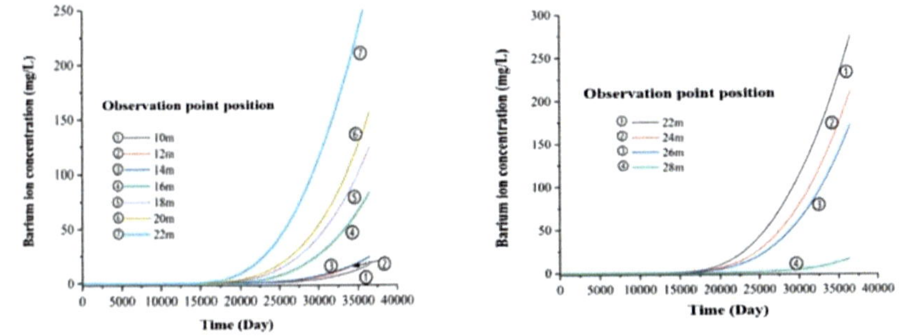

(a) Ba^{2+} concentration map at observation points 10 m to 22 m

(b) Ba^{2+} concentration map at observation points 22 m to 28 m

Fig. 14.4 Barium ion concentration map at the observation point. **a** Ba^{2+} concentration map at observation points 10–22 m, **b** Ba^{2+} concentration map at observation points 22–28 m

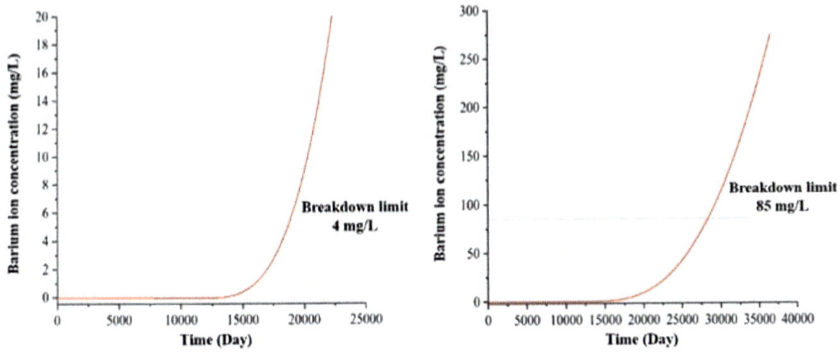

(a) Breakdown curve 1

(b) Breakdown curve 2

Fig. 14.5 Barium ion breakdown curve. **a** Breakdown curve 1, **b** Breakdown curve 2

In order to evaluate the cutoff effect of the cutoff barrier, the data of the observation point with the fastest Ba^{2+} migration speed (22 m) was selected here to calculate the time of Ba^{2+} breakdown barrier. The so-called leachate breakdown of the cutoff barrier refers to the phenomenon that the leachate passes through the cutoff barrier after a series of processes such as long-term penetration, migration, dispersion and adsorption, resulting in the failure of the cutoff barrier. The failure here refers to the loss of the effectiveness of the cutoff barrier to protect the environment, that is, the time node when the leachate harms the environment. According to the allowable landfill control limit of hazardous waste given in the "Hazardous Waste Landfill Pollution Control Standard" (GB18598-2019) (Ministry of Ecological Environment 2019), the target limit of barium is 85 mg/L. According to the groundwater index and limit given in the Groundwater Quality Standard (GB/T 14848-2017) (General Administration of Quality Supervision, Inspection and Quarantine of the People's

Republic of China, National Standardization Management Committee 2017), the target limit for barium is 4 mg/L.

According to the target barium limit of 4 mg/L, the breakdown time of plastic concrete is 49.91 years. According to the target limit of barium of 85 mg/L, the breakdown time of plastic concrete is calculated to be 77.73 years. It can be concluded that the breakdown time calculated according to different Ba^{2+} limits is different, and the larger the breakdown limit, the longer the breakdown time. In order to be safe, the limit of Ba^{2+} concentration 4 mg/L given in groundwater was selected as the breakdown standard.

14.5 Conclusions

By using Hydrus-2D finite element simulation software to explore the migration rule of Ba^{2+} in the vertical plastic concrete cutoff barrier, the following conclusions are drawn:

1. The vertical plastic concrete cutoff barrier has a strong blocking effect on Ba^{2+} migration. After a long period of infiltration, only a small amount of Ba^{2+} passed through the cutoff barrier driven by the water flow.
2. 19 observation points are set on the outer side of the vertical plastic concrete cutoff barrier from the top down, 2 m apart. The pollutant migration at different time was obtained by finite element calculation. According to the data, Ba^{2+} migrates the fastest in the vertical plastic concrete cutoff barrier at 22 m, which is the weakest place of the cutoff barrier and the most easily broken down by Ba^{2+}.
3. According to the target limit of barium of 4 mg/L, the breakdown time of plastic concrete is 49.91 years. According to the target limit of barium of 85 mg/L, the breakdown time of plastic concrete is calculated to be 77.73 years. The breakdown time calculated according to different Ba^{2+} limits is different, and the larger the breakdown limit, the longer the breakdown time. For safety reasons, a Ba^{2+} concentration of 4 mg/L is used as the breakdown standard in engineering.

Acknowledgements The authors would like to thank National Natural Science Foundation of China (Grant No. 42007263), the 'Qing Lan Project' of the Jiangsu Higher Education Institutions of China (2023), the Key Research and Development Program (Social Development) project of Zhenjiang (Grant No. SH2022017), the Science and Technology Project of the Ministry of Housing and Urban-Rural Development of China (Grant No. 2019-K-136), the China Postdoctoral Science Foundation funded project (Grant No. 2020M671297) and the Project of Research Innovation of Graduate Student in Jiangsu Province of China (Grant No. SJCX23_2230) for supporting the research.

References

Carsel RF, Parrish RS (1988) Developing joint probability distributions of soil water retention characteristics. Water Resour Res 24(5):755–769

Chang DM, Jiang YT et al (2022) Application of plastic concrete impervious wall technology in Xincheng regulation and storage reservoir project. Henan Water Resour South-to-North Water Diversion 51(6):45–47

Chen HH, Chen HW, He JT, et al (2006) Health-based risk assessment of contaminated sites principles and methods. Earth Sci Front 216–223

Dong SJ, Cai TZ, et al (2023) Study on the migration rule of petroleum hydrocarbon pollutants in soil of petrochemical enterprises based on hydrus-1d model. Chem Eng Manage 35–37

Fan M (2020) Environmental impact assessment analysis of barite mining project. Energy Conserv Environ Prot 42–43

Fu Z, Zhang FC, et al (2021) Study on simulation and disposal of soil pollution in reservoir based on Hydrus model. HaiHe Water Resour 80–83

General Administration of Quality Supervision, Inspection and Quarantine of the People's Republic of China, National Standardization Management Committee. Groundwater quality Standard (GB/T 14848–2017). China Standards Press, Beijing (2017)

La ZR (2023) Application of plastic concrete wall in seepage prevention and reinforcement construction of earth dam. Heilongjiang Hydraul Sci Technol 51(9):105–107+114

Li GB (2023) Study on performance of plastic concrete based on different siliceous admixture. Heilongjiang Hydraul Sci Technol 51(6):1–4

Liu Y, Wei ZC, Li MY et al (2021) Research progress of barite separation process and resource overview. Conserv Utiliz Miner Resour 41(6):117–123

Luo J, Liu BH et al (2023) Heavy metal migration in contaminated soil simulated by using Hydraus-1D software. J Univ Jinan (Sci Technol) 37(04):444–448

Lv JJ, Shi LJ, et al (2022) Study on modified soil infiltration system model for wastewater treatment based on Hydrus-1D. Environ Prot Technol 28(6):1–5+16

Ministry of Ecological Environment (2019) Hazardous waste landfill pollution control standards (GB 18598-2019). China Environmental Science Press, Beijing

Peranić J, Arbanas Ž, Cuomo S, et al (2018) Soil-water characteristic curve of residual soil from a flysch rock mass. Geofluids

Qian XD, Zhu W, Xu HQ et al (2017) Design and construction of anti-pollution barriers for landfills and contaminated sites (part I). Science Press, Beijing

Simunek J, Sejna M, Van Genuchten MT (2006) The HYDRUS software package for simulating two-and three-dimensional movement of water, heat, and multiple solutes in variably-saturated media, user manual, version 1.0. University of California-Riverside Research Reports

Van Genuchten MT (1980) A closed-form equation for predicting the hydraulic conductivity of unsaturated soil. Soil Sci Soc Am J 44(5):892–898

Wang HT (2008) Contaminant migration kinetics in porous media. Higher Education Press, Beijing

Wang XH, Yan XH, Li X (2016) Research on pollution characteristics and resource utilization risk of barium slag. J Environ Eng Technol 6(2):170–174

Wang GW (2023) Influence of geopolymer on the early strength and microstructure of plastic concrete. Concrete 122–125

Wang JY, Zhang HL, Hou CH, et al (2023) Present situation and development trend of barite industry technology in China. Ceramics 29–32

Wu JZ (2018) Study on the permeability resistance of plastic concrete. Water Resour Plan Des 107–111

Wu JF (2022) Study on performance of concrete cut-off wall of water supply project in pearl in river delta. Water Conservancy Sci Technol Econ 28(9):135–138

Xie HJ (2008) A study on contaminant transport in layered media and the performance of landfill liner systems. Zhejiang University

Xu B, Zhao LQ, Hu YJ, et al (2021) Analysis of heavy metal migration mechanism during road backfill of MSWI bottom ash as based on geostudio simulation software. In: 2021 Science and technology annual meeting of Chinese society of environmental sciences. Industrial Building Magazine Co., LTD, Tianjin

Yu FP (2023) Analysis of the application of plastic concrete impermeable wall in the reinforcement project of reservoir in front of the rock. Ground Water 45(4):287–289

Zhang P (2005) Experimental study on the material properties of plastic concrete and its application. Zhengzhou University

Zheng LS (2022) Construction technology of plastic concrete cut-off wall in reservoir engineering. Henan Water Resour South-to-North Water Diversion 51(12):113–114

Chapter 15
Evaluation and Perspective of the Thermal Treatment Technologies of Medical Waste for Energy and Value-Added Products

Muhammad Usman, Aijun Li, Yongda Huang, Tong Zhang, Yuhang Zheng, Shuai Li, and Hong Yao

Abstract Medical Waste management poses a considerable challenge due to its hazardous nature and environmental concerns associated with its disposal in the healthcare industry. Thermal treatment technologies are sustainable and viable methods for the disposal of medical waste along with value-added energy products. This paper focuses on thermal treatment technologies such as incineration, pyrolysis, and gasification, and their potential to convert medical waste into energy and value-added by-products. Firstly, the current challenges and complexities in the thermal treatment technologies of medical waste were critically examined, and issues related to efficiency, regulatory compliance, emissions, environmental impact, and cost and economic viability were especially highlighted. Secondly, this analysis elucidated the regulations and innovative solutions to address these challenges. Finally, this evaluation concluded by offering perspectives on advancements in the field regarding technological innovations, resource recovery, energy generation, and environmental sustainability along with future challenges to waste reduction, technological integration, regulatory evolution, climate and health, and emerging waste streams paving for the sustainable management of medical waste and future research. In this exploration, it is aimed to provide insights into thermal treatment technologies for policymakers, environmentalists, and researchers working to address the critical issue of medical waste disposal while promoting sustainability and resource recovery.

M. Usman · A. Li (✉)
China-EU Institute for Clean and Renewable Energy, Huazhong University of Science and Technology, Wuhan 430074, China
e-mail: hust_laj@hust.edu.cn

A. Li · Y. Huang · T. Zhang · Y. Zheng · S. Li · H. Yao
School of Energy and Power Engineering, Huazhong University of Science and Technology, Wuhan 430074, China

Y. Huang · S. Li · H. Yao
State Key Laboratory of Coal Combustion, Huazhong University of Science and Technology, Wuhan 430074, China

Z. Wan (ed.), *Water Resources Management and Water Pollution Control*, Environmental Science and Engineering, https://doi.org/10.1007/978-3-031-53456-0_15

Keywords Medical Waste · Waste Disposal · Thermal Treatment Technologies · Resource Recovery · Environmental Sustainability

15.1 Introduction

One of the most emerging issues in developed and developing countries that have received much attention is rapid increases in waste generation and inefficient handling, creating many socioeconomic and environmental issues (Yousefian et al. 2020). Medical waste can be defined as waste generated at healthcare institutions such as emergency medical settings, hospitals, dispensaries, and medical laboratories, consisting of substances and pathogenic microbes that exhibit detrimental effects on both human health and the environment. Medical waste can be categorized as infectious or non-infectious waste, 85% of medical waste is non-infectious. Medical waste generated by healthcare facilities, regardless of their specific concentration and purpose, consists of materials like textiles, paper, polymer rubber, food and biological compounds, antiseptics, glass, gypsum, and metals. The components show distinct variations in physical, chemical, and mechanical characteristics (Khaskhachikh et al. 2021). Millions of people in underdeveloped nations who work with medical waste are at risk of contracting HIV, hepatitis B, and hepatitis C because of the infectious nature of the waste (Singh et al. 2022). The healthcare sector is the fifth largest emitter of greenhouse gases, and its growth is rapid, estimated at 20% per annum. The increase in medical waste generation is due to the rapid increase in population, healthcare awareness, better quality of life, utilization of non-reusable packaging, advances in medical technology, and subsequently growth of the healthcare industry. The COVID-19 pandemic caused a significant increase in both environmental pollution and public health crises. Medical waste contains organic or organic components with a calorific value ranging from 2 to 40 MJ/kg, making them a good option for energy or value-added products recovery (Mazzei and Specchia 2023). However, the inefficient management and severe environmental and human health risks make medical waste an imperative to explore waste treatment sustainably.

Thermal treatment technologies such as incineration, pyrolysis, and gasification, as shown in Fig. 15.1, are promising avenues to address medical waste disposal. These treatment methods can potentially treat and transform medical waste into valuable energy and resources. These treatment methods encompass many processes that can safely and efficiently manage medical waste. Thermal treatment technologies offer the advantage of reducing the volume of medical waste along with energy and by-products, making them an environmentally friendly alternative to medical waste disposal (Purnomo et al. 2021). This paper will focus on the potential of various thermal treatment methods to convert medical waste into energy and value by-products. This investigation aims to conduct an in-depth examination of these technologies, exposing positive and negative impacts as well as their long-term durability and effectiveness in the field of medical waste management. In this study, current challenges in the thermal treatment of medical waste, such as emissions, cost, regulations, and resource recovery, along with prospects and future challenges, will be discussed.

15.2 Thermal Treatment Technologies for Medical Waste

15.2.1 Incineration

Incineration refers to the method of waste disposal that involves the combustion of waste materials at high temperatures within furnaces. The procedure, as mentioned earlier, involves the extraction of dangerous substances, as well as the reduction of waste bulk and volume, ultimately transforming it into ash, which is harmless. Incineration works best with trash that is 60% flammable or less. Pathological and infectious trash, as well as sharp waste, are amenable to incineration (Shareefdeen 2012). The temperatures involved are quite high, ranging from about 980 to almost 2000 °C. To ensure full combustion, organic materials are burned in the presence of high concentrations of oxygen. The primary benefit is reducing material volume by as much as 90%. Ash, flue gases, particulate matter, and heat are the primary byproducts of burning. It's a steady process that effectively removes contaminants and sterilizes surfaces.

Although this technology is ideal for all forms of infectious medical waste, it should be avoided when dealing with pressurized reactive chemical waste, gas containers, silver salts, batteries, reactive chemical waste, heavy metals, PVC plastics, sealed ampoules, vials, radioactive materials, and unstable pharmaceuticals. Bottom slag and fly ash are produced as hazardous by-products of incineration because of they contain a high concentration of poisonous metals. Incineration also generates other harmful gases such as hydrogen chloride, hydrogen fluoride, and sulfur dioxide, necessitating the installation of an exhaust gas purification system (Mazzei and Specchia 2023). Polychlorinated dioxins and furans are examples of molecules that could be released during the process that are both toxic and carcinogenic. Textiles, medical supplies, and plastics all have very high dioxin levels (Jiang et al. 2019). Dioxin generation is affected by many factors, including the type of waste input, the temperature and duration of combustion, the amount of chlorine in the flue gas, the presence of heavy metals, and the amount of carbon that remains in the fly ash. The presence of chlorine from plastics is the primary cause of high dioxin emissions from waste incinerators, as plastics make up the bulk of medical waste. However, the main cause of such emissions is sloppy combustion management. To ensure the safety of the surrounding area and the purity of the air, medical waste incinerators need to have their emissions closely monitored. Damage to health and the environment, as well as technical issues such as tube corrosion, can result from inadequate management of these factors (Yu et al. 2015). Flue gas treatment is necessary to prevent all of these emissions from medical waste combustion. Air pollution treatment accounts for over half of an incinerator's operating cost. Common

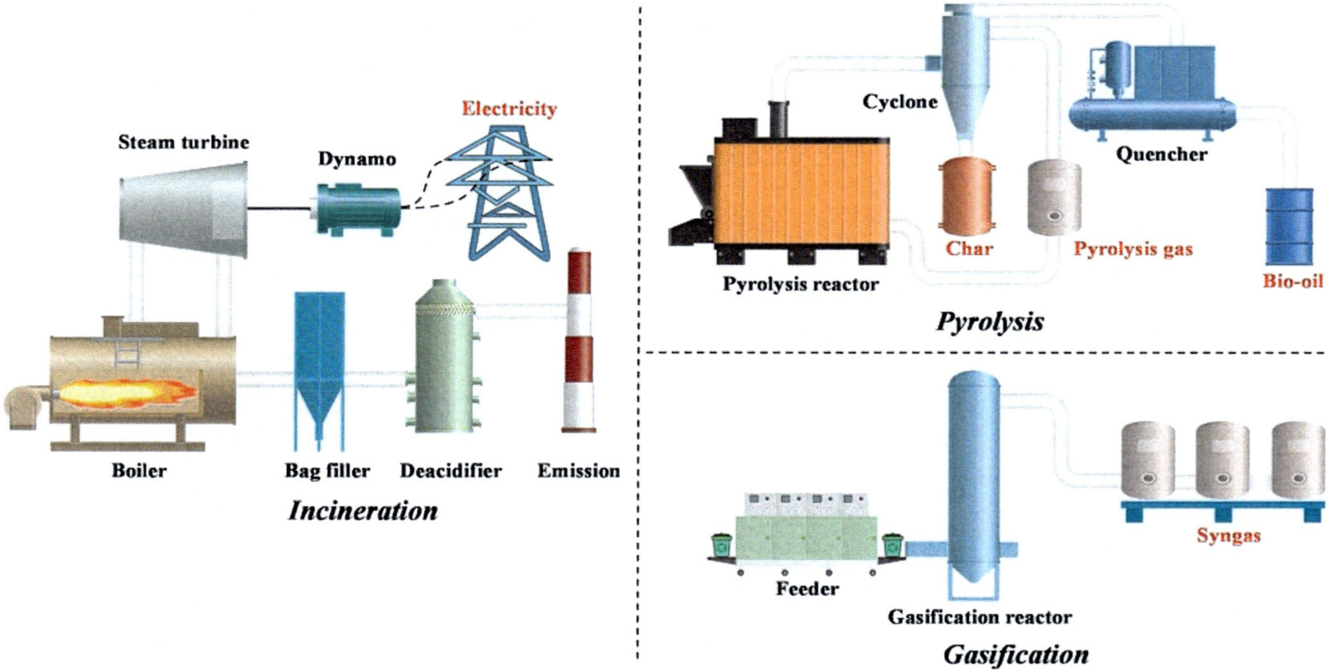

Fig. 15.1 Diagrams and schemes of thermal treatment technologies of medical waste (incineration, pyrolysis, and gasification)

components of a gas cleaner unit include a quencher for the exhaust, a water or alkali scrubber, a catalytic converter, a carbon filter, and a bag filter to capture the finer particles. Before the flue gas can be discharged into the atmosphere via a tall stack, it must be treated by the equipment. Combustion heat, which may be used to create steam and electricity, is the only usable byproduct. Ash is a byproduct of this technology that needs to be analyzed for any trace amounts of heavy metals (Purnomo et al. 2021).

15.2.2 Pyrolysis

In the absence of oxygen, medical waste is degraded by an endothermic process known as pyrolysis. Depending on the feedstock and processing conditions, different amounts of char, bio-oil, and pyrolysis gas will be produced. When a catalyst is used, the process is referred to as catalytic pyrolysis rather than the more common thermal pyrolysis. Catalysts are used to fix problems including low-quality liquid oil, dirty fuel gas, and excessive energy needs. Based on the rate of heating, this process can be classified as slow (270–630 °C) pyrolysis and residence time of minutes, rapid (580–980 °C) pyrolysis and residence time of below 3 s, or flash (780–1030 °C) pyrolysis and residence time of 0.03–1.5 s (Mazzei and Specchia 2023). Pyrolysis is a fascinating technique for valorizing polymeric wastes since it has several benefits. It can firstly create large amounts of liquid products throughout a broad temperature range. Moreover, the method is adaptable in generating valuable products as it allows for the optimization of process parameters like temperature, pressure, and residence time to yield products with specific characteristics. These items have qualities that make them appropriate for use as chemical feedstock in addition to fuels. Fossil fuels can be replaced by bio-oil, which is produced through the pyrolysis of medical waste. In various fields, Biochar is widely used as an anode material, catalyst, adsorbent, and photocatalytic support. Furthermore, it is an environmentally favorable process because it produces less carbon dioxide (1.0–9.1 vol%) and carbon monoxide (0.8–3.9 vol%) than burning (Chen et al. 2014). The pyrolysis treatment method of medical waste, when combined with nontoxic municipal solid waste, presents an alternative strategy for the diversification of medical waste treatment. The pyrolysis technique offers adequate economics, low secondary pollutants, and a high rate of energy recovery. An almost entirely automated equipment set for medical waste pyrolysis with simultaneous gas retrieval requires limited space, the control of standard equipment, industrial function, minimal modifications in the market, and effective marketing.

As previously indicated, pyrolysis technology may achieve energy circulation, lower energy consumption, lower processing costs, and achieve economic viability by using the gas produced by medical waste treatment (Giakoumakis et al. 2021). All types of medical waste can be processed via pyrolysis, but the process is costly because of the high energy consumption, maintenance needs, and pretreatments involved. Furthermore, it is difficult to produce steady combustion, which produces harmful gases such as PAHs, HCl, SO_2, and NO_x, endangering both ecological security and public health (Datta et al. 2018). Moreover, a second gas-cleaning system can be used to stop the release of harmful gases (Isahak et al. 2012). The chlorine used to create HCl comes primarily from PVC or physiological saline found in medical waste. Gaseous pollution from HCl during pyrolysis of medical waste could lead to acid rain and corrosion. Bio-oil's HCl can cause high acidity, low stability, HHV, petrol engine damage, and a considerable value reduction (Hunsinger et al. 2002).

15.2.3 Gasification

Through the application of high heat in a carefully managed atmosphere, carbon-based materials can be gasified to produce a range of gases including carbon monoxide, hydrogen, carbon dioxide, methane, and those with longer chains of hydrocarbons. Energy costs can be reduced and product yields can be controlled with the help of a catalyst. Depending on their chemical makeup, gasification byproducts can be used as fuel or as chemical feedstock (syngas). Gasification is an appealing option for revalorizing medical waste, which mainly contains plastic waste, because of the possibility of creating energy, energy carriers, and chemicals from syngas generated (Sansaniwal et al. 2017). The first stage is drying in which moisture content is drastically decreased at a modest temperature (about 100–200 °C), the second is pyrolysis during this stage, tar and gaseous fractions are generated, the third stage is combustion during the combustion stage these products of pyrolysis are decomposed, and fourth is reduction in which the products of combustion are decomposed, resulting in a gaseous mixture of smaller molecules; and during the gasification stage, the final syngas are generated (Ramos et al. 2018). Process integration with existing power production equipment, such as the steam cycle, gas turbines, and gas engines, makes it possible to implement this technology in a power plant. In addition to minimizing the release of harmful byproducts including NO_x, SO_x, and heavy metals, gasification also generates some ash, vitrified slag, and gaseous discharge as byproducts (Bosmans et al. 2013). The oxygen-deficient environment in the gasifier does not favor the generation of dioxins and furans since they need adequate oxygen to be created or re-formed. Gasification is an attractive option for revalorizing plastic waste, the main element of medical waste, due to the possibility of producing electricity and chemicals from syngas.

However, a major drawback of plastic gasification is the presence of tar in the final gas product. High working temperatures need expensive running costs and construction materials that can withstand those temperatures, making this a complex and pricey facility. The components of synthesis gas are carbon monoxide, carbon dioxide, hydrogen, chlorofluorocarbons, higher hydrocarbons, nitrogen, and trace amounts of other contaminants. As a heterogeneous feedstock, medical waste results in syngas with a higher proportion of incombustible mixes and a lower LHV than syngas produced from biomass (Brems et al. 2013). The elementary composition of the waste, the LHV of the waste, the amount of the injected oxidant, the nature of the gasifying agent, the reactor pressure, the temperature gradient within the reactor, the posttreatment of the gas obtained, and the gasifier design are the primary determinants of the performance and composition (Mazzei and Specchia 2023).

As shown in Table 15.1, the three kinds of thermal treatment technologies of medical waste (incineration, pyrolysis and gasification) own different advantages, disadvantages, and challenges. Incineration can be an efficient way for medical disposal but the initial cost and emissions and ash heavy ash metal possess certain limitations that need to be addressed. Pyrolysis offers a promising solution for the environmentally friendly and sustainable way of medical waste management. This technique addresses the concerns of energy generation, resource recovery, volume reduction, and pathogen destruction while we are facing issues like initial investment, complexity, public participation, and regulatory compliance. Pyrolysis can become a key player in the future due to its potential waste management. In comparison, gasification is a sustainable way to dispose of medical waste and it solves issues related to pyrolysis and incineration.

15.3 Current Challenges in the Thermal Treatment of Medical Waste

If we look for challenges in thermal treatment technologies that are previously discussed to destroy the pathogen, reduce volume, and minimize infection spread risk they have certain concerns and the first one is emissions, controlling the emission is a key challenge to the emission of heavy metals, dioxins, furans, and PAH which is necessary to address these issues of emission control monitoring. These emissions rates can be as high as 40,000 times the Stockholm Convention's emission limits. To address the controlling of the emissions, a cleaning system is required, which consists of different systems like catalytic converters, bag filters, and scrubbers, which are necessary for environmental sustainability. Another technique required to control emissions is to change parameters such as temperature, feed rate, and residence time which will help to control emissions (Purnomo et al. 2021; Calin 2011). For controlling emissions requires a lot of cost that will increase the cost of operation and this cost is 50% of the operation cost. Secondly, health and environmental regulations are difficult to address and must be adhered to and met with regulatory compliance. Every region has its regulations for emissions and these emissions standards also affect the cost of operation and add complexity. So, to meet this issue collaboration with the waste management institutions, health care institutions, and government bodies can help in streamlining the regulatory landscape (Tan et al. 2015).

Maximizing the valuable resource recovery, such as clean ash and metals from the treatment of medical waste is an important aspect that needs to be addressed. Balance generation of energy along with resource recovery is a difficult task. Metals recovered during the treatment of medical waste can be recycled and utilized again (Li et al. 2020). If these metals are not recovered, this is a waste of resources and also harmful

Table 15.1 Advantages, disadvantages, and challenges among incineration, pyrolysis and gasification of medical waste

Treatment methods	Advantages	Disadvantages	Challenges
Incineration	– Great flexibility of materials all types of waste can be burned – No requirement for pre-treatment for disinfection – Low consumption of feedstock due to high CV of medical waste – Destruction of pathogens – Volume reduction of waste	– Recoverable product is heat that can be utilized – Harmful and corrosive gas production – Emission of high carbon – Regular maintenance and skill labor requirements	– Treatment technology for flue gas – Storage and capture technology for CO_2
Pyrolysis	– Wide range of chemical feedstock production – Carbon emission is low – Product generation is flexible according to the condition – Very suitable for PP and PE – The oxygen-free environment prevents dioxins and furan production – No disinfect treatment required	– Not suitable for PVC, PET, and rubber – Air emissions – Combustible gases raise security concerns controls are needed for pollutants – High investment cost – Pre-treatment and skilled labor requirement	– Maturity in technology development
Gasification	– Suitable for PP and PE – Able to treat PVC – No requirement for the disinfection stage – Emissions are low – High energy efficiency	– Strict controls are needed as combustible gases – A complex and expensive facility – A high amount of energy requirement – Tar formation is high	– Removal of tar technology needed – Efficient heat recovery system integration

to the environment and health. Large frameworks should include transportation, import taxes, site preparation, installation, commissioning, project management, and administrative costs in their capital budgets. Labor, utilities, supplies, maintenance, and recurring validation tests are all included in operating costs. The cost of medical waste management is as high as 5 times of municipal solid waste (Lee et al. 2004). A large amount of waste generated in the world is not properly managed, and almost 40% of the laborers working in the medical waste management and treatment sector are injured during their work. This is due to a lack of training in the people working in this sector (Singh et al. 2022).

15.4 Prospects and Future Challenges

Research and development that are going on for the thermal treatment of medical waste advanced technologies such as fluidized thermal treatment technologies and rotary kiln incineration are being developed to improve the efficiency of the systems. Optimizing the conditions, enhancing energy recovery, and reducing emissions are the aims of innovations. Control and monitoring of systems are now becoming sophisticated to alter the operational conditions and waste composition. These innovations are dealing with the current challenges and provide a sustainable and environmentally friendly approach to medical waste management (Olan et al. 2023). The circular economy is a concept that is attracting momentum in the context of the treatment of medical waste. Advanced sorting out technologies such as robotic and NIR spectroscopy are being utilized to recover resources. These innovative technologies enable the efficient segregation of valuable materials, including plastic, metals, and clean ash, from medical waste treatment. The recovered resources can be utilized in different applications in industry, and these resources will promote sustainability and minimize the need for virgin resources (Kheirabadi and Sheikhi 2022). Shifting towards a circular economy aligns not only with objections of the environment but also offers beneficial economic benefits by creating revenues and minimizing disposal expenses. Integration with the existing energy generation systems thermal treatment technologies possesses certain benefits of not only reducing initial cost but also making the existing systems more efficient and beneficial.

If we look at future challenges, the first thing that we need to take care of is the reduction of medical generation by optimizing the supply chain management, sterilizing technique, and good relations between health facilities, manufacturers and regulators to implement reduction measures. Transportation and storage should efficiently take place by segregating hazard and non-hazard waste separately. Containers should be covered and labeled properly, and staff working in this stream should be well-trained. Integration of the current technologies with existing infrastructure can be a complex task, and new considerations for developed technologies, such as worker training and adoption of new procedures are crucial. So, a lot of care is required for planning and investment to upgrade the system and accommodate the invocations while efficiently maintaining the existing system. Public participation and engagement in the newly developed technologies should also be considered for acceptance. Also, open and clear communication regarding health and the environment is necessary (Giakoumakis et al. 2021). The regulatory system for dealing with medical waste is dynamic and prone to change. It is essential for facilities to be knowledgeable about new standards and requirements and to adjust their operating methods to conform to these advances to effectively deal with the ever-evolving nature of regulatory changes. Keeping up with the ever-changing pollution standards, waste disposal regulations, and safety regulations is essential. Ensuring conformity to regulatory requirements while simultaneously anticipating and adapting to forthcoming regulatory changes is of the highest importance (Padmanabhan and Barik 2019).

15.5 Conclusion

A medical waste issue is crucial for both developed and developing countries. Inefficient treatment poses serious health and environmental effects, and medical waste can be managed sustainably by incineration, pyrolysis, and gasification. Incineration reduces volume and kills microorganism; however, complex handling of byproducts, ash, and flue gas is an issue. Pyrolysis is an effective technique for the destruction of pathogens and recovering valuable resources like bio-oil, pyrolysis gas, and char, but concerns like high investment cost, pre-treatment and combustible gas, and emission management. Gasification can produce energy from medical waste, especially polymers. This technique solves the problem of incineration and pyrolysis but requires facility design and careful combustion gas management and facility design. Thermal treatment technologies for medical waste require emission control, resource recovery, and regulatory compliance. Emerging technologies include real-time monitoring and enhanced combustion. Resource recovery and sustainability are promoted via circular economy growth. Waste reduction, technological integration, public participation, legislation, global collaboration, climate change mitigation, and new waste streams are future issues. To overcome these difficulties, healthcare providers, regulators, and the public must collaborate. These technologies' current and future issues require innovation, teamwork, and careful compliance with laws and best practices. Health and the environment can benefit from sustainable medical waste management.

Acknowledgements The authors gratefully acknowledge the financial supports provided by the National Natural Science Foundation of China (52076092, 52220105006) and China Postdoctoral Science Foundation (2023M731181).

References

Bosmans A, Vanderreydt I, Geysen D et al (2013) The crucial role of waste-to-energy technologies in enhanced landfill mining: a technology review. J Clean Prod 55:10–23

Brems A, Dewil R, Baeyens J et al (2013) Gasification of plastic waste as waste-to-energy or waste-to-syngas recovery route. Nat Sci (irvine) 5(6):695–704

Calin G (2011) Report of the special rapporteur on the adverse effects of the movement and dumping of toxic and dangerous products and wastes on the enjoyment of human rights, the United Nations General Assembly.

Chen D, Yin L, Wang H et al (2014) Pyrolysis technologies for municipal solid waste: a review. Waste Manag 34(12):2466–2486

Datta P, Mohi G, Chander J (2018) Biomedical waste management in India: critical appraisal. J Lab Physicians 10(1):6–14
Giakoumakis G, Politi D, Sidiras D (2021) Medical waste treatment technologies for energy, fuels, and materials production: a review. Energies (Basel, Switz.) 14(23):8065
Hunsinger H, Jay K, Vehlow J (2002) Formation and destruction of PCDD/F inside a grate furnace. Chemosphere 46(9–10):1263–1272
Isahak WNRW, Hisham MWM, Yarmo MA et al (2012) A review on bio-oil production from biomass by using pyrolysis method. Renew Sust Energ Rev 16(8):5910–5923
Jiang X, Li Y, Yan J (2019) Hazardous waste incineration in a rotary kiln: a review. Waste Disposal and Sustainable Energy 1(1):3–37
Khaskhachikh VV, Kornil'eva VF, Gerasimov GY (2021) Investigation into the Pyrolysis of Medical Waste in a fixed-bed Reactor. J Eng Phys Thermophy 94(3):580–586
Kheirabadi S, Sheikhi A (2022) Recent advances and challenges in recycling and reusing biomedical materials. Curr Opin Green Sustain Chem 38:100695
Lee BK, Ellenbecker MJ, Moure-Ersaso R (2004) Alternatives for treatment and disposal cost reduction of regulated medical wastes. Waste Manag 24(2):143–151
Li YM, Wang CF, Wang LJ et al (2020) Removal of heavy metals in medical waste incineration fly ash by Na_2EDTA combined with zero-valent iron and recycle of Na_2EDTA: acolumnar experiment study. J Air Waste Manage Assoc 70(9):904–914
Mazzei HG, Specchia S (2023) Latest insights on technologies for the treatment of solid medical waste: a review. J Environ Chem Eng 11(2)
Olan S, Padhye LP, Kumar M et al (2023) Review on distribution, fate, and management of potentially toxic elements in incinerated medical wastes. Environ Pollut 321:121080
Padmanabhan KK, Barik D (2019) Health hazards of medical waste and its disposal. Energy from toxic organic waste for heat and power generation. Woodhead Publishing, Cambridge
Purnomo CW, Kurniawan W, Aziz M (2021) Technological review on thermochemical conversion of COVID-19-related medical wastes. Resour Conserv Recycl 167
Ramos A, Monteiro E, Silva V et al (2018) Co-gasification and recent developments on waste-to-energy conversion: a review. Renew Sust Energ Rev 81:380–398
Sansaniwal SK, Pal K, Rosen MA et al (2017) Recent advances in the development of biomass gasification technology: a comprehensive review. Renew Sust Energ Rev 72:363–384
Shareefdeen ZM (2012) Medical waste management and control. J Environ Prot 3(12):1625–1628
Singh N, Ogunseitan OA, Tang Y (2022) Medical waste: current challenges and future opportunities for sustainable management. Crit Rev Environ Sci Technol 52(11):2000–2022
Tan ST, Ho WS, Hashim H et al (2015) Energy, economic and environmental (3E) analysis of waste-to-energy (WTE) strategies for municipal solid waste (MSW) management in Malaysia. Energy Convers Manag 102:111–120
Yousefian F, Hassanvand MS, Nodehi RN et al (2020) The concentration of BTEX compounds and health risk assessment in municipal solid waste facilities and urban areas. Environ Res 191:110068
Yu J, Qiao Y, Jin L et al (2015) Removal of toxic and alkali/alkaline earth metals during co-thermal treatment of two types of MSWI fly ashes in China. Waste Manag 46:287–297

Chapter 16
Application of BP Neural Network in Pyrolysis Treatment of Organic Solid Waste

Yuhang Zheng, Aijun Li, Yongda Huang, Tong Zhang, Muhammad Usman, Nanxi Bie, and Hong Yao

Abstract The improper management of organic solid waste can precipitate many environmental issues. Pyrolysis, characterized by its capacity for rapid reduction and resource utilization of organic solid waste, has garnered considerable attention. Machine learning, an autonomous system capable of knowledge acquisition and integration, holds promise for predicting the thermochemical conversion products of organic matter. The backpropagation (BP) neural network, owing to its simplistic model structure, high prediction accuracy, and robust self-learning capability, has seen extensive application in the predictive modeling of organic solid waste pyrolysis treatment in recent years. Despite the breadth of research, a comprehensive summary of the BP neural network's research outcomes in organic solid waste pyrolysis and a comparative analysis of the models are conspicuously absent. This study systematically overviews the application of BP neural network in organic solid waste pyrolysis. Initially, the specific structure of the BP neural network model and parameter optimization are delineated. Subsequently, an evaluation of the BP neural network's practical application in organic solid waste pyrolysis treatment is provided, along with a comparison of the parameter selection across different models. Finally, the efficacy of the BP neural network model in predicting the scenario of organic solid waste pyrolysis treatment is expounded upon, and potential directions for future development are discussed.

Y. Zheng · A. Li (✉) · Y. Huang · T. Zhang · N. Bie · H. Yao
School of Energy and Power Engineering, Huazhong University of Science and Technology, Wuhan 430074, China
e-mail: hust_laj@hust.edu.cn

A. Li · M. Usman
China-EU Institute for Clean and Renewable Energy, Huazhong University of Science and Technology, Wuhan 430074, China

Y. Huang · N. Bie · H. Yao
State Key Laboratory of Coal Combustion, Huazhong University of Science and Technology, Wuhan 430074, China

Z. Wan (ed.), *Water Resources Management and Water Pollution Control*, Environmental Science and Engineering, https://doi.org/10.1007/978-3-031-53456-0_16

Keywords Organic Solid Waste · Pyrolysis · Machine Learning · BP Neural Networks

16.1 Introduction

Organic solid waste encompasses the solid and semi-solid organic pollutants generated by human activities, including agricultural waste (such as straw, livestock and poultry waste, and peels), urban waste (such as domestic waste and sludge), and industrial waste (such as industrial sludge and industrial waste residue). China has long grappled with a substantial output and diverse range of organic waste. Improper treatment methods for organic solid waste can precipitate environmental issues, underscoring the urgency of recycling organic solid waste. Common recycling methods for organic solid waste include classification recovery, landfill, composting, anaerobic digestion, incineration, pyrolysis, and gasification. Pyrolysis, in particular, has garnered considerable attention due to its ability to rapidly achieve organic solid waste reduction and resource utilization (Lee and Lee 2016; Ciavatta et al. 1993; Mata-Alvarez et al. 2000; Nandhini et al. 2022).

Given the multitude of factors influencing the composition of pyrolysis products and potential synergistic effects among various factors, it is impractical to exhaustively conduct experimental study on the pyrolysis products of organic solid waste under all composition ratios and working conditions (Fassinou et al. 2009; Yang et al. 2022). Linear weighting alone can easily lead to significant errors. The artificial neural network (ANN) is an intelligent computing model that emulates the structure and function of human brain neurons. It can learn the complex mapping relationship between multi-dimensional input and output variables by training many existing sample data, thereby establishing a mathematical model based on sample data (Fan et al. 2018). The neural network does not require prior knowledge of the specific physical or chemical mechanisms and the mathematical relationship between the input and output parameters in the reaction system (Ascher et al. 2022). It exhibits robust accuracy and resilience in data mining. Therefore, it holds promise for predicting the thermochemical conversion products of organic matter. As a typical artificial neural network model, the backpropagation (BP) neural network has been extensively applied in organic solid waste pyrolysis due to its simple model structure, high prediction accuracy, and strong self-learning ability (Sun et al. 2016).

Despite some studies exploring the utilization of the BP neural network in organic solid waste pyrolysis, a systematic summary of these research results is conspicuously absent. This study aims to provide a systematic overview of the utilization of the BP neural network in the pyrolysis of organic solid waste. It summarizes the efficacy of the BP neural network in predicting the pyrolysis treatment of organic solid waste and discusses the differences in model optimization.

16.2 The Structure and Optimization of BP Neural Network

16.2.1 Structure of BP Neural Network

BP neural network model is a multi-layered feedforward network trained via the error backpropagation algorithm. The diagram of BP neural network structure is displayed in Fig. 16.1. The input layer serves as the information input terminal, while the output layer represents the prediction target. The hidden layer functions to receive, process, and transmit information from either the input or output layers. The count of layers and neurons within each layer significantly influences the network's training accuracy and generalization capability. Studies suggest that, given an appropriate count of neurons in the hidden layer, a BP neural network with just a single hidden layer can effectively estimate the relevant data (Deng et al. 2021).

The core concept of the BP neural network is the input, transformation, and output of information. Initially, external information is received by the neurons of the input layer, then passed on to the hidden layer for processing. The neurons in the hidden layer reconstruct the input information and then pass the results to the output layer. In this process, based on the output results of forward propagation, we can calculate the error. This error will guide us to modify the connection weights and thresholds through backpropagation (Otchere et al. 2021). By continuously adjusting these parameters, we can make the actual output results as close as possible to our expected output results. According to the primary principle, the calculation process of the algorithm is shown as follows (Xie 2022):

16.2.1.1 Forward Propagation of the Signal

The input net_i of the i th node of the hidden layer is shown in Eq. (16.1):

$$net_i = \sum_{j=1}^{10} \omega_{ij} x_j + \theta_i \tag{16.1}$$

where, x_j denotes the input of the jth node of the input layer, $j = 1, 2, \ldots, 10$; ω_{ij} denotes the weight value between the ith node of the hidden layer and the jth node of the input layer; θ_i is the threshold of the ith node of the hidden layer.

The output o_i of the ith node in the hidden layer is shown in Eq. (16.2):

$$o_i = \phi(net_i = \phi\left(\sum_{j=1}^{10} \omega_{ij} x_j + \theta_i\right) \tag{16.2}$$

where ϕ denotes the activation function of the hidden layer.

The input net_k of the kth node in the output layer is shown in Eq. (16.3):

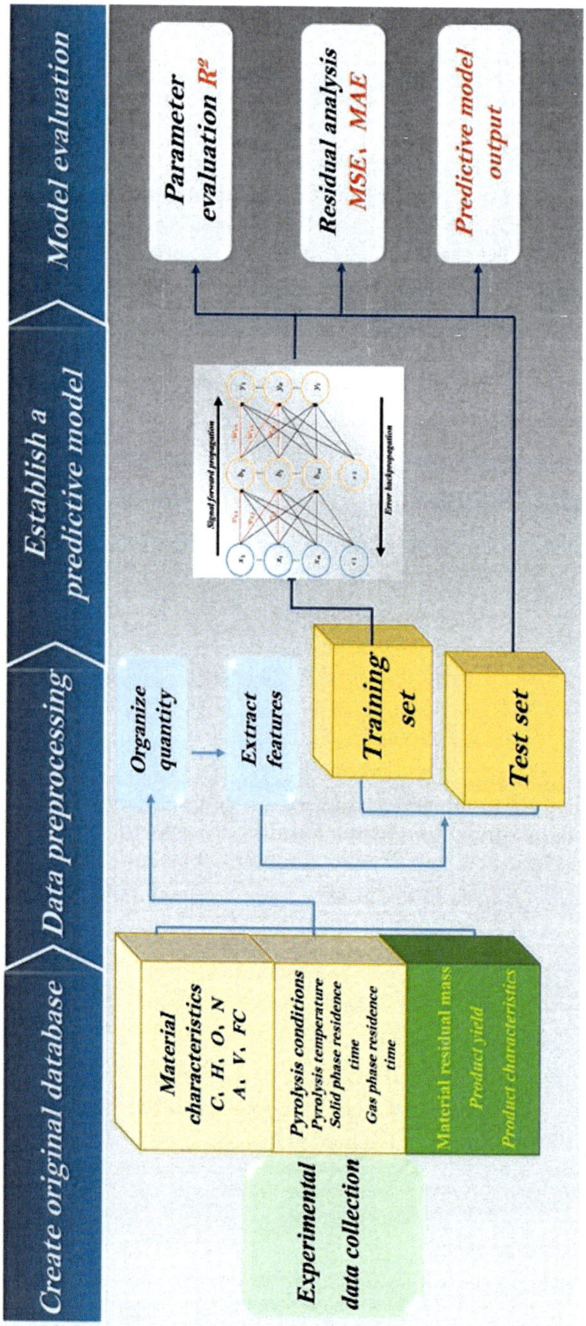

Fig. 16.1 The diagram of BP neural network structure

$$net_k = \sum_{i=1}^{h} \omega_{ki} y_i + a_k = \sum_{i=1}^{h} \omega_{ki}\, \phi\left(\sum_{j=1}^{10} \omega_{ij} x_j + \theta_i\right) + a_k \tag{16.3}$$

Here, ω_{ki} denotes the weight value between the kth node of the output layer and the ith node of the hidden layer, $i = 1, 2, \ldots, h$; a_k denotes the threshold of the kth node in the output layer, where $k = 1, 2, 3$.

The output o_k of the kth node of the output layer is shown in Eq. (16.4):

$$o_k = \psi(net_k) = \psi\left(\sum_{i=1}^{h} \omega_{ki} y_i + a_k\right) = \psi\left[\sum_{i=1}^{h} \omega_{ki} \phi\left(\sum_{j=1}^{10} \omega_{ij} x_j + \theta_i\right) + a_k\right] \tag{16.4}$$

where ψ denotes the activation function of the output layer.

16.2.1.2 Backpropagation of Error

The backpropagation of error is based on the output error of neurons in each layer, and the weight and threshold of each layer are corrected layer by layer from the output layer. Therefore, the final output result of the network is close to the anticipated value.

The error function E_p of each sample P is shown in Eq. (16.5):

$$E_p = \frac{1}{2} \sum_{k=1}^{3} (T_k - o_k)^2 \tag{16.5}$$

where T_k is the expected output of the kth node in the output layer.

The total error function E_P of the network for P training samples is shown in Eq. (16.6):

$$E_P = \frac{1}{2} \sum_{p=1}^{P} \sum_{k=1}^{3} (T_k - o_k)^2 \tag{16.6}$$

On the basis of the gradient descent method, the thresholds and weights of the hidden and output layers are then corrected. The correction calculation formula is shown in Eqs. (16.7)–(16.10):

$$\Delta\omega_{ki} = -\eta \frac{\partial E}{\partial \omega_{ki}} \tag{16.7}$$

$$\Delta a_k = -\eta \frac{\partial E}{\partial a_k} \tag{16.8}$$

$$\Delta\omega_{ij} = -\eta \frac{\partial E}{\partial \omega_{ij}} \tag{16.9}$$

$$\Delta\theta_i = -\eta \frac{\partial E}{\partial \theta_i} \tag{16.10}$$

The BP neural network continuously repeats the above two processes in the learning process until the error meets the requirements or reaches the maximum number of iterations. Finally, a network model with a better prediction effect is obtained.

16.2.2 Model Parameters Optimization

When building and training BP neural network, the selection of transfer function, number of hidden layer neurons and learning rate will affect BP neural network's training cost and prediction effect. Parameter optimization is a crucial step in machine learning and deep learning, entailing the selection of the optimal combination of parameters to enhance the performance of the model. In the context of Backpropagation (BP) neural networks, prevalent parameter optimization methods include the Levenberg–Marquardt (LM), Gradient Descent, Particle Swarm Optimization (PSO) (Yang 2023), among others.

The core component of the neural network, known as the transfer function, establishes the correlation between input and output variables. The commonly used transfer functions are Logsig logarithmic S-type transfer function, Tansig tangent S-type transfer function and Purelin linear transfer function. The three transfer functions are expressed in Eqs. (16.11)–(16.13) (Hu et al. 2022).

$$\text{Logsig: } f(x) = \frac{1}{1 + e^{-x}} \tag{16.11}$$

$$\text{Tansig: } f(x) = \frac{2}{1 + e^{-2x}} - 1 \tag{16.12}$$

$$\text{Purelin: } f(x) = x \tag{16.13}$$

According to the Eqs. (16.11)–(16.13), the Logsig function maps the input interval of the neuron from $(-\infty, +\infty)$ to (0,1), the Tansig function maps the input interval of the neuron from $(-\infty, +\infty)$ to $(-1,1)$, and the Purelin function retains the input of the neuron.

The count of neurons in the hidden layer determines the structure, which is the focus of attention in neural network research. Suppose the count of neurons in the hidden layer is too Insufficient. In that case, the model cannot play the role of nonlinear fitting, so it cannot accurately reflect the correlation between input and output. When the number of hidden layer neurons is too large, the too complex network structure will increase the time cost of training, and it is easy to over-fit the sample data to reduce the network's generalization ability (Zhao et al. 2022b). There is no set guideline for choosing the quantity of neurons in the hidden layer. The most widely empirical formula is shown in Eq. (16.14) (Zhao et al. 2022a).

$$h = \sqrt{i + o} + a \tag{16.14}$$

where h, i and o is the count of neurons in hidden layer, input neurons, and output neurons, accordingly, and a is an arbitrary constant between 0 and 10.

The learning rate refers to the step size of each iteration of the model in the course of training, representing the weight and the amplitude of the threshold update. When the learning rate is overly large, the weight and threshold change span are large, and the network convergence speed is fast. Still, the gradient has the potential to vacillate around the minimum value or even lack convergence, while too small a learning rate will make the network convergence process very slow and easy to get trapped in the local optimum (Li et al. 2022b). Therefore, although it can be arbitrarily valued between 0 and 1 according to the definition, learners of neural networks tend to tend to set the learning rate between 0.001 and 0.1.

16.3 Application and Model Comparison of BP Neural Network

16.3.1 Predicting Residual Mass of Materials

At present, more and more scholars use the BP neural network to explore the pyrolysis process parameters of organic solid waste based on thermogravimetric data to obtain the residual mass of the material. Li et al. (2022a) predicted the pyrolysis mass loss of several biomasses, including pine wood chips, cow dung, cowpea stalks, and bamboo, using the pyrolysis temperature as the input variable. The model has a good predictive effect ($R^2 > 0.998$). Selvarajoo et al. (2020) constructed a BP neural network for weight loss prediction of banana peels during the pyrolysis process. Mayol et al. (2018) used the data of thermo gravimetric analysis (TGA) and BP neural network data to predict algae pyrolysis kinetic data. The R^2 of the model was greater than 0.97.

16.3.2 Predicting the Yield of Pyrolysis Products

The prediction of three-phase products of organic solid waste pyrolysis is also frequently reported. Chen et al. (2018) used carbon, hydrogen, nitrogen, oxygen, moisture, ash, granularity of raw materials, temperature, and fluidization number as inputs to predict the thermal value of bio-oil and the quantity of biomass pyrolysis products. The R^2 of the model's test set ranged between 0.800 and 0.932. Merdun and Sezgin (2018) constructed models for the pyrolysis variables and product yields (including biochar, bio-oil, and gas mixture) of 11 different biomasses. The models achieved prediction accuracies with R2values of 0.91, 0.96, and 0.95 for biochar, bio-oil, and gas mixture respectively throughout the training phase. Aydinli et al. (2017) made predictions about the output of solid, liquid, and gaseous products during the pyrolysis process of cotton, tea, olives, and hazelnuts. The regression coefficient R^2 of the model reached 0.99. From the above, it suggests that the BP neural network has achieved more success in predicting biomass thermal conversion yield.

16.3.3 Predicting the Characteristics of Pyrolysis Products

Some scholars utilized the BP neural network to forecast the properties of pyrolysis products from organic solid waste. Liao et al. (2019) constructed a highly accurate neural network model ($R^2 > 0.9$) that is capable of effectively predicting the overall production and specific surface dimension of activated carbon generated by the pyrolysis of various biomass raw materials. Sun et al. (2016) used operating temperature, air speed, and biomass particle size as inputs to emulate and estimate the preference of gas in the pyrolysis products of waste biomass. Zhao et al. (2022b) modeled the pyrolysis process of rice straw and pine wood, successfully predicting the distribution of their pyrolysis products. And they optimized the model through algorithms, achieving good predictive performance ($R^2 = 0.99238$).

16.3.4 Comparison of the Models Parameters

As described in the second section, the transfer function of the hidden layer and the node count are structural parameters of the neural network. The pace of neural network training is dependent on the learning rate. The selection of these elements is fundamental to the comprehensive training cycle of the neural network.

Table 16.1 illustrates that, based on each transfer function's characteristics, Logsig or Tansig is typically chosen as the of hidden layer neurons, and Purelin is used as the transfer function of output layer neurons in the prediction of thermochemical conversion products of organic solid waste. This combination has demonstrated excellent predictive outcomes in most studies. The count of hidden layer neurons is usually dictated by referring to Eq. (16.14) to establish the range, and then the optimal number of hidden layer neurons is obtained through multiple experiments. The learning rate, like the number of hidden layer neurons, does not strictly adhere to existing rules, and its adjustment requires experiments to assess the performance of network models with different learning rates.

Moreover, as shown in the Table 16.1, the Levenberg–Marquardt (LM) method is predominantly employed for supervised training in the optimization of the neural network algorithm for biomass thermochemical conversion. Some researchers also

Table 16.1 Comparative analysis of models describing the pyrolysis treatment of organic solid waste using BP neural networks

Types of organic solid waste	Data sources	Number of hidden layers	Number of hidden layer neurons	Learning rate	Training algorithm	Hidden layer transfer function	Output layer transfer function	Model inputs	Model outputs	Accuracy	References
Pine sawdust, cattle dung, kidney bean stalk, bamboo	–	1	4	–	LM	–	–	Pyrolysis temperature	Mass loss	$R^2 > 0.998$	Li et al. (2022a)
Banana peels	88	1	21	–	LM	Tansig	Purelin	Pyrolysis time, temperature	Mass loss	$R^2 = 0.999$	Selvarajoo et al. (2020)
Algal mat	–	1	100	0.001	LM	Tansig	–	Instantaneous temperature, target temperature, heating rate	Mass loss	$R^2 > 0.97$	Mayol et al. (2018)
Bio-oil	166, 175, 71	1	8	0.1	Trainbr	Tansig	Purelin	The performance of biomass raw materials	Three-phase yield	R^2: 0.800–0.932	Chen et al. (2018)
Various	72–44	2	10	–	LM	Logsig	–	Moisture, volatile, fixed carbon, ash, HHV, heating rate, Temperature, content of C, H, O, N	Three-phase yield	RMSE = 5.71–9.16	Merdun and Sezgin (2018)

(continued)

Table 16.1 (continued)

Types of organic solid waste	Data sources	Number of hidden layers	Number of hidden layer neurons	Learning rate	Training algorithm	Hidden layer transfer function	Output layer transfer function	Model inputs	Model outputs	Accuracy	References
Cotton, tea, olives and hazelnuts	18	1	9	–	Gradient descent with adaptive learning	Tansig	Tansig	Fixed carbon, volatile, moisture and ash content, lignin, cellulose, hemicellulose, temperature	Three-phase yield	$R^2 = 0.99$	Aydinli et al. (2017)
Various	168–20	1	9	–	LM	Tansig	Purelin	Fixed carbon, volatile and ash content, carbonization temperature and time, activation temperature and time, content of C, H, O	Yield of activated carbon and BET surface area	$R^2 > 0.92$	Liao et al. (2019)

(continued)

Table 16.1 (continued)

Types of organic solid waste	Data sources	Number of hidden layers	Number of hidden layer neurons	Learning rate	Training algorithm	Hidden layer transfer function	Output layer transfer function	Model inputs	Model outputs	Accuracy	References
Pine sawdust	14	1	7	–	LM	Logsig	Purelin	Operating temperature, space velocity and biomass particle size	Selectivity of H_2, CO, CH_4, CO_2	$R^2 = 0.9904$	Sun et al. (2016)
Rice straw and pine wood	45	2	15	–	PSO	Logsig	Purelin	Approximate analysis, limit analysis, pyrolysis temperature and oxygen concentration	The yields of pyrolysis products (water, tar, gas and carbon) and two gases (CO and CO_2) were calculated	$R^2 = 0.99238$	Zhao et al. (2022b)

use Particle Swarm Optimization (PSO), Bayesian regularization (trainbr), and gradient descent and adaptive learning to enhance the training of the model.

16.4 Conclusion and Future Outlooks

Employing the BP neural network as a predictive tool for all facets of the pyrolysis process can serve as a valuable resource in addressing real-world challenges. As mentioned above, the BP neural network exhibits significant promise in modeling the pyrolysis process of organic solid waste, as well as the production and characteristics of pyrolysis products. The application mainly focuses on predicting material residual mass, production of three-phase products, and pyrolysis product characteristics and has achieved good prediction results. This is of great significance to developing organic solid waste pyrolysis treatment technology.

However, although the BP neural network has many successful application cases in predicting the organic solid waste pyrolysis process, these attempts are mainly focused on biomass. Given the similar composition of biomass materials and the relatively close spread of pyrolysis products, intricacy of model training and forecasting is considerably reduced. Predicting the pyrolysis process of organic solid waste with complex components is still challenging. In addition, the selection of input features greatly influences the prediction effect of neural networks. However, the current research often selects input variables based on experience, and there are relatively few reports on the correlation investigation of input variables before constructing neural networks. To enhance the efficiency of network computation and the precision of predictions, future prediction processes could employ methods such as grey relational analysis to sequentially select the input features of the BP neural network based on their degree of correlation. Finally, further considering a more comprehensive range of training algorithms to adjust the hyperparameters more systematically is crucial for optimizing the neural network. Developing a more comprehensive neural network model to improve the design and optimization of the organic solid waste pyrolysis process will become an important direction for future research on organic solid waste treatment.

Acknowledgements The authors gratefully acknowledge the financial supports provided by the National Natural Science Foundation of China (52076092, 52220105006), National Key Research and Development Plan (2019YFC1906604) and China Postdoctoral Science Foundation (2023M731181).

References

Ascher S, Watson I, You S (2022) Machine learning methods for modelling the gasification and pyrolysis of biomass and waste. Renew Sustain Energy Rev 155:111902

Aydinli B, Caglar A, Pekol S et al (2017) The prediction of potential energy and matter production from biomass pyrolysis with artificial neural network. Energy Explor Exploit 35(6):698–712

Chen X, Zhang H, Song Y et al (2018) Prediction of product distribution and bio-oil heating value of biomass fast pyrolysis. Chem Eng Process Process Intensif 130:36–42

Ciavatta C, Govi M, Pasotti L et al (1993) Changes in organic matter during stabilization of compost from municipal solid wastes. Biores Technol 43(2):141–145

Deng Y, Zhou X, Shen J et al (2021) New methods based on back propagation (BP) and radial basis function (RBF) artificial neural networks (ANNs) for predicting the occurrence of haloketones in tap water. Sci Total Environ 772:145534

Fan M, Hu J, Cao R et al (2018) A review on experimental design for pollutants removal in water treatment with the aid of artificial intelligence. Chemosphere 200:330–343

Fassinou WF, Steene LVD, Toure S et al (2009) Pyrolysis of pinus pinaster in a two-stage gasifier: influence of processing parameters and thermal cracking of tar. Fuel Process Technol 90(01):75–90

Hu Z, Yuan Y, Li X et al (2022) Yield prediction of "thermal-dissolution based carbon enrichment" treatment on biomass wastes through coupled model of artificial neural network and AdaBoost. Biores Technol 343:126083

Lee NH, Lee CY (2016) The effect of solid waste landfill method on decomposition of pollutants in semi-aerobic landfill structure. J Korea Organ Res Recycl Assoc 8(4):153–159

Li J, Yao X, Ge J et al (2022a) Investigation on the pyrolysis process, products characteristics and BP neural network modelling of pine sawdust, cattle dung, kidney bean stalk and bamboo. Process Saf Environ Prot 162:752–764

Li Z, Huang J, Wang J et al (2022b) Comparative study of meta-heuristic algorithms for reactor fuel reloading optimization based on the developed BP-ANN calculation method. Ann Nucl Energy 165:108685

Liao M, Kelley SS, Yao Y (2019) Artificial neural network-based modeling for the prediction of yield and surface area of activated carbon from biomass. Biofuels Bioprod Biorefining 13(4):1015–1027

Mata-Alvarez J, Macé S, Llabrés P (2000) Anaerobic digestion of organic solid wastes. An overview of research achievements and perspective. Biores Technol 74(1):3–16

Mayol AP, Maningo JMZ, Chua-Unsu AGAY et al (2018) Application of artificial neural networks in prediction of pyrolysis behavior for algal mat (LABLAB) biomass. In: 2018 IEEE 10th International Conference on Humanoid, Nanotechnology, Information Technology, Communication and Control, Environment and Management (HNICEM). IEEE, Heidelberg, pp 1–13

Merdun H, Sezgin IV (2018) Modeling of pyrolysis product yields by artificial neural networks. Int J Renew Energy Res 8(2):1178–1188

Nandhini R, Berslin D, Sivaprakash B et al (2022) Thermochemical conversion of municipal solid waste into energy and hydrogen: a review. Environ Chem Lett 20(3):1645–1669

Otchere DA, Ganat TOA, Gholami R et al (2021) Application of supervised machine learning paradigms in the prediction of petroleum reservoir properties: comparative analysis of ANN and SVM models. J Petrol Sci Eng 200:108182

Selvarajoo A, Muhammad D, Arumugasamy SK (2020) An experimental and modelling approach to produce biochar from banana peels through pyrolysis as potential renewable energy resources. Model Earth Syst Environ 6:115–128

Sun Y, Liu L, Wang Q et al (2016) Pyrolysis products from industrial waste bio-mass based on a neural network model. J Anal Appl Pyrol 120:94–102

Xie T (2022) Study on the Oxidative pyrolysis and combustion characteristics and prediction models of typical biomass. Ph.D. Thesis, University of Science and Technology of China, Hefei

Yang H, Liu Y, Bai G et al (2022) Study on the co-pyrolysis characteristics of oil-based drill cuttings and lees. Biomass Bioenerg 160:106436

Yang TH (2023) Research on neural network architecture search and parameter optimization method based on particle swarm optimization algorithm. North China University of Technology

Zhao L, Zhang M, Wang H et al (2022a) Monitoring of free fatty acid content in mixed frying oils by means of LF-NMR and NIR combined with BP-ANN. Food Control 133:108599

Zhao S, Xu W, Chen L (2022b) The modeling and products prediction for biomass oxidative pyrolysis based on PSO-ANN method: an artificial intelligence algorithm approach. Fuel 312:122966

Chapter 17
Evaluating Main Odor Sources and Health Risk Assessment of Volatile Compounds in the Whole Process of Municipal Solid Waste Treatment

Qiuxia Wei, Jianjun Cai, Mengnan Ma, Longxian Su, and Long Huang

Abstract Municipal solid waste (MSW) incineration can contaminate the environment, leading to the "NIMBY effect" and negatively impacting society. MSW reduction, harmlessness, and recycling are critical issues in urban development. This study focused on a incineration and analyzed the odor pollutants. The concentration and composition of odor pollutants varied based on the incineration process. The highest total mass concentration of odorous pollutants was found in the closed garbage pit (127.90 mg/m^3), followed by the leachate pool (61.95 mg/m^3) and discharge area (47.04 mg/m^3). The feed pit was identified as the primary source of odor emissions (60.65%), followed by the unloading area (18.59%) and leachate pool (13.35%). Health risk assessment revealed that risks were concentrated in the leachate pool, discharge area, and feed pit, with the leachate tank posing the highest health risk (HI 11.72, LCR 8.9×10^{-3}). The cumulative threshold dilution factor of malodorous gas in the entire garbage disposal process was 48,167.51.

Keywords Municipal solid waste · Incineration · Malodorous gas · Cumulative threshold dilution multiple · Health risk assessment

Supplementary Information The online version contains supplementary material available at https://doi.org/10.1007/978-3-031-53456-0_17.

Q. Wei · J. Cai (✉) · M. Ma · L. Su · L. Huang
School of Architecture and Transportation Engineering, Guilin University of Electronic Technology, Guilin, China
e-mail: caijianjun@guet.edu.cn

J. Cai
National Engineering Laboratory for High-Efficiency Recovery of Refractory Nonferrous Metals, School of Metallurgy and Environment, Central South University, Changsha, China

Z. Wan (ed.), *Water Resources Management and Water Pollution Control*, Environmental Science and Engineering, https://doi.org/10.1007/978-3-031-53456-0_17

17.1 Introduction

China's rapid economic development and urban population expansion have led to an increase in the types and quantities of garbage generated in daily life, resulting in a significant rise in municipal solid waste (MSW) output across the country (Cai et al. 2015; Guo et al. 2017).

The main methods of MSW disposal worldwide are incineration (Rand et al. 2000; Liu et al. 2006), landfill (Crawford and Smith 2016; Kamaruddin et al. 2017) and composting (Kumar 2011; Hargreaves et al. 2008). In China, the volume of MSW removal and transportation in 2021 was 287 million tons, with a 99.9% harmless treatment rate. The country has 1407 domestic waste treatment facilities, including 542 landfills, 583 incinerators, and 282 other treatment facilities. The harmless treatment capacity of these facilities is about 248 million tons per year, with incineration gradually becoming the primary disposal method (Li et al. 2016).

Analyzing malodorous gas produced during each process of waste incineration can help prevent intermittent odor pollution and solve the 'NIMBY' problem. Many scholars have studied the malodorous gases produced during the incineration of different municipal wastes and their impact on the environment and human health risks (Nie et al. 2018; Nanda and Berruti 2021a; Wu et al. 2020; Liu and Zheng 2020). However, the release characteristics and health risks of volatile compounds during the entire process of MSW collection and processing in China are still unclear (Xu et al. 2020; Titto and Savino 2019; Domingo et al. 2020; Wu et al. 2018; Fang et al. 2012). Therefore, a study was conducted in a waste-to-energy plant.

The incineration facility covers an area of 98,000.5 m^2 and can process 1,500 tons of domestic waste per day, or 547,500 tons per year. The process generates a large amount of pollutants, including slag (Lin 2006), fly ash (Zhang et al. 2020, 2021), volatile compounds (VC) (Nanda and Berruti 2021a), volatile organic compounds (VOC) (Komilis et al. 2004; Nanda and Berruti 2021b; Zhang et al. 2013) and other conventional pollutants. These pollutants pose significant risks to the local atmospheric environment and the health of nearby residents (Vinti et al. 2021). Therefore, it is crucial to analyze the environmental impact of air pollutants produced by domestic waste incineration plants.

17.2 Experimental Materials and Methods

17.2.1 Sampling

The samples analyzed in this study were collected from a domestic waste incineration power plant, including gas samples from the unloading area (UA), waste pit (FA), slag pile (SS), leachate pond (LP), and flue gas purification system (FG). Samples were collected using 6L silanized Suma tanks in accordance with the technical specification for environmental monitoring of odor (HJ905-2017) and analyzed within

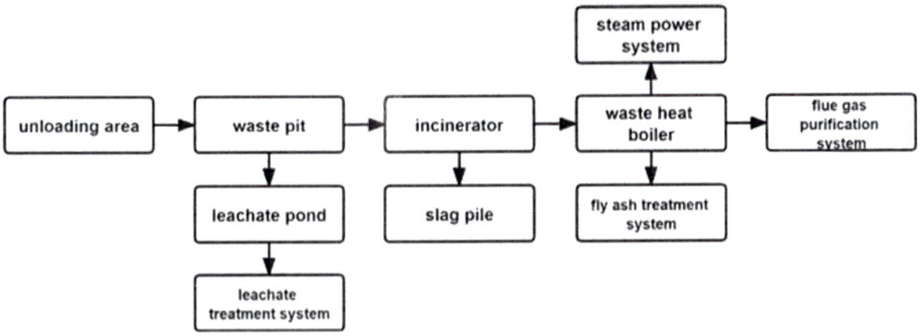

Fig. 17.1 Process flow of waste incinerator

1 week. The unloading area, leachate pond, and slag pile were semi-enclosed, while the waste pit was closed (see Fig. 17.1 for the process flow).

17.2.2 Experimental Method

The gas sample was concentrated in a pre-concentrator using a rotary valve at 80 °C and a transfer line at 100 °C. The sample volume was 400 mL, and the primary cold trap was set at −150 °C with a preheating temperature of 20 °C, a desorption temperature of 120 °C, and a baking temperature of 140 °C for 5 min. The secondary cold trap was set at −20 °C with a preheating temperature of 20 °C, a desorption temperature of 180 °C, and a baking temperature of 190 °C for 5 min. The tertiary cold trap was set at −160 °C with a desorption temperature of 150 °C. The injection time was 3 min, the baking time was 2 min, and the waiting time was 32 min.

After pre-concentration, the gas sample was analyzed using gas chromatography-mass spectrometry (GC–MS). The chromatographic column was DB-5MS (60 m × 0.32 mm × 1.0 μm), and the carrier gas was ammonia (purity >99.999%) at a flow rate of 1.5 mL/min. Non-split injection was used, with an injection port temperature of 180 °C. The temperature program for the column oven was 30 °C (held for 5 min), followed by a ramp to 150 °C at 5 °C/min, and then a ramp to 220 °C at 15 °C/min (held for 10 min). Mass spectrometry was performed using an ion source temperature of 230 °C, quadrupole temperature of 150 °C, interface temperature of 280 °C, and ionization energy of 70 eV. Full scan mode was used for qualitative analysis, and selected ion mode was used for quantitative analysis. A total of 120 volatile organic standard substances and internal standards were purchased from Spectra Gases Inc.

17.2.3 *Analysis of Odor Pollutants*

17.2.3.1 Normalization Processing

The total concentration of malodorous gas in each region and the mass concentration of various gas components were normalized using the following formula:

$$M_i = \frac{A_i - A_{min}}{A_{max} - A_{min}} A_i$$

where A_i is a certain component of a certain type of malodorous gas, A_{min} is the gas with the lowest mass concentration in a certain type of malodorous gas, A_{max} is the gas with the highest mass concentration in a certain type of malodorous gas, and M_i is the normalization process of malodorous gas A_i result.

Similarly, the mass concentration of each type of gas in each area is normalized, and the processing formula is as follows:

$$Y_i = \frac{B_i - B_{min}}{B_{max} - B_{min}}$$

In the formula, B_i is the total mass concentration of a certain type of malodorous gas, B_{min} is the gas with the lowest mass concentration in a certain type of malodorous gas, B_{max} is the type of gas with the highest mass concentration in a certain type of malodorous gas, and Y_i is the value of the malodorous gas B_i Normalize the results. Finally, according to M_i and Y_i, the final integral C_i of a certain type of a certain component of malodorous gas is obtained, and the specific formula is as follows:

$$C_i = M_i \times Y_i$$

At this time, according to the magnitude of the C_i integral value, we can preliminarily search for the odorous gas with greater influence.

17.2.3.2 Screening of Key Odor Pollutants

According to the following formula, calculate the threshold dilution factor of a single odor pollutant in different technological processes.

$$D_i = \frac{c_i}{c_{i,T}}$$

In the formula: D_i is the threshold dilution multiple of odor pollutant i; c_i is the mass concentration of odor pollutant i, mg/m^3; $c_{i,T}$ is the odor threshold of odor pollutant i, mg/m^3.

$$OU_T = \sum D_i$$

17.2.3.3 Health Risk Assessment of Odor Pollutants

The inhalation reference doses for non-carcinogens and carcinogenic slope factors for carcinogens, as established by the US Environmental Protection Agency, were used to assess the non-carcinogenic risk (HI) and carcinogenic risk (LCR) associated with malodorous pollutants.

$$HI = \frac{c_i \times IR \times ET \times EF \times ED}{BW \times AT \times RFC}$$

$$LCR = \frac{c_i \times IR \times ET \times EF \times ED \times SF}{BW \times AT}$$

The non-carcinogenic risk hazard index (HI) was calculated using the formula provided, with IR representing the respiration rate (0.66 m^3/h), ET representing the daily exposure time (8 h/d), EF representing the continuous exposure frequency in one year (250d/a), ED representing the duration of exposure (25a), BW representing body mass (65 kg), and AT representing the average life expectancy of the population. The inhalation reference dose of non-carcinogens (RFC) and carcinogenic slope factor (SF) of carcinogens were obtained from the Integrated Risk Information System (IRIS) of the US Environmental Protection Agency. Lifetime cancer risk (LCR) was also calculated using the provided formula, with AT set at 25,550 days (70a) for carcinogenic risk assessment and 9125 days (25a) for non-carcinogenic risk assessment.

17.3 Results and Discussion

17.3.1 Composition of Odor Pollutants in Different Regions

Figure 17.2a shows the concentration distribution of malodorous gas components in the garbage pit. Oxygen-containing compounds (51.62 mg/m^3), ammonia (41.96 mg/m^3), and sulfur-containing compounds (10.32 mg/m^3) were the most prevalent. Acetaldehyde and hydrogen sulfide were the most concentrated oxygenated and sulfur-containing compounds, respectively. Figure 17.2b shows the concentration distribution of malodorous gas components in the leachate pool. Oxygenated compounds (26.27 mg/m^3), ammonia (13.00 mg/m^3), and sulfur compounds (8.08 mg/m^3) were the most prevalent. Ethyl acetate and carbon disulfide were the most concentrated oxygenated and sulfur-containing compounds, respectively.

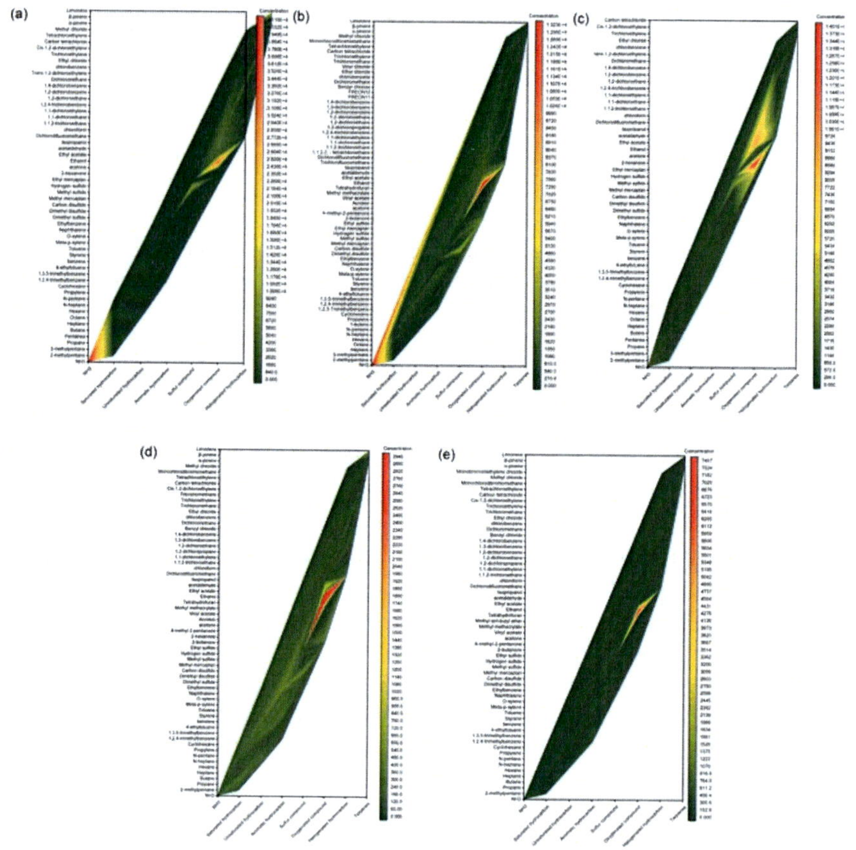

Fig. 17.2 The mass concentration (ug/m^3) of odorous pollutants in different technological processes: **a** garbage pit; **b** leachate pool; **c** unloading area; **d** flue gas purification area; **e** slag dump

Figure 17.2c shows the concentration distribution of malodorous gas components in the garbage unloading area. Oxygen-containing compounds (32.49 mg/m^3), halogenated hydrocarbon compounds (5.25 mg/m^3), and sulfur-containing compounds (3.09 mg/m^3) were the most prevalent. Ethanol, 1,1-dichloroethane, and hydrogen sulfide were the most concentrated oxygenated, halogenated hydrocarbon, and sulfur-containing compounds, respectively. Figure 17.2d shows the concentration distribution of malodorous gas components in the flue gas purification area. Oxygenated compounds (18.10 mg/m^3) were the most prevalent malodorous gases. Figure 17.2e shows the concentration distribution of malodorous gas components in the slag stacking area. Oxygenated compounds (8.18 mg/m^3) were the most prevalent malodorous gases, while other types of malodorous gases had a very low proportion due to their destruction and degradation during high-temperature incineration.

17.3.2 *Weight Integral of Odorous Gas Components in Garbage in Different Areas*

Figure 17.3 is the weight integral distribution map of the garbage odor gas components in different areas. As shown in the figure, compared with other areas, the total mass concentration of odorous gases in garbage pits is the largest, so the classification of odorous gases in this area and the weight integral of each component under each category are significantly higher than the weight integral of odor gas components in other regions. Judging from the final weight integral, ammonia, pentane, propylene, 1,2,4-trimethylbenzene, ethanol, methylene chloride, and limonene located in the feed pit have the highest weight integral and are the main odorous gases. Secondly, the area with higher weight integral of malodorous gas is the leachate pool, and the weight integral of malodorous gas located in this area is higher. Among them, the main malodorous gases with weight integral of 0.6 include: hexane, cyclohexane, **p-Xylene**, O-Xylene, Carbon Disulfide, Trichlorofluoromethane, Limonene, Ethanol.

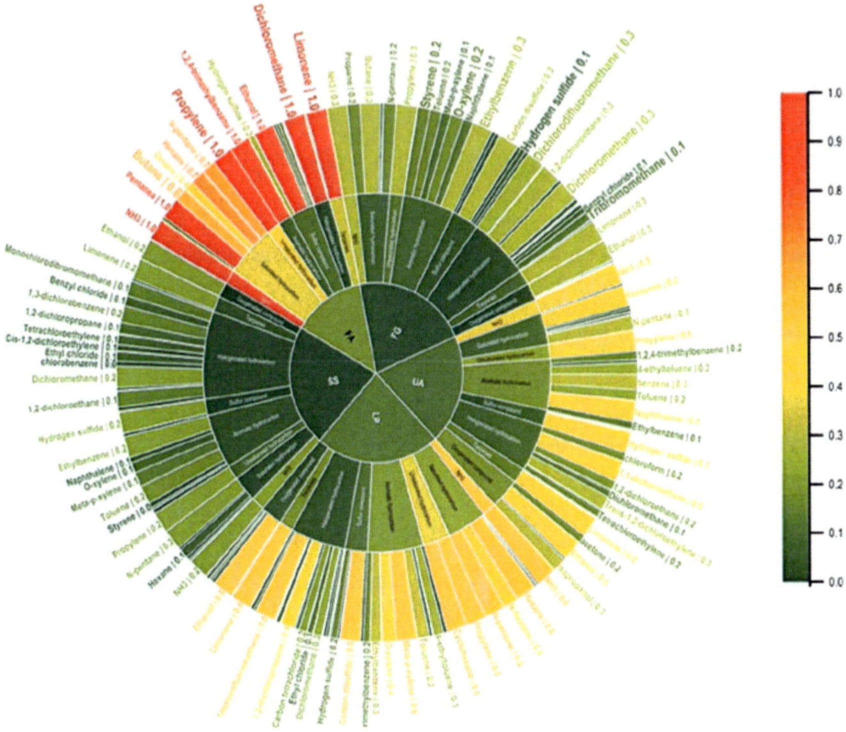

Fig. 17.3 Weight integral distribution map of garbage odor gas components in different regions

17.3.3 The Threshold Dilution Factor of Each Component of Malodorous Gas in Each Interval

Table 17.1 shows the threshold dilution factor of malodorous gas in the garbage pit. It can be seen from the table from the perspective of categories, the threshold dilution factor of sulfur-containing malodorous gases is the largest, and its main source is hydrogen sulfide gas, up to 24,328.19; followed by oxygen-containing malodorous gases, the cumulative dilution threshold is 3277.85. The cumulative threshold dilution factor of aldehyde is the largest, which is 3205.624.

Table S1 presents the threshold dilution factors for malodorous gases in the leachate pond. The cumulative threshold dilution factor of malodorous gas in the leachate area is 6434.63, accounting for only 22.03% of the cumulative dilution factor of odorous gas threshold in the feed pit. The main types of malodorous gases are sulfur-containing compounds (80.43%) and oxygen-containing malodorous gases (16.74%). Hydrogen sulfide, methyl mercaptan, and acetaldehyde are the main malodorous gases that need to be controlled.

Table S2 presents the threshold dilution factors for malodorous gases in the unloading area. The cumulative threshold dilution factor of malodorous gas in the unloading area is 8952.01, accounting for 30.64% of the cumulative dilution factor of odorous gas threshold in the feed pit. The main malodorous gas is sulfur-containing compounds (84.70%), with hydrogen sulfide being the primary malodorous gas, followed by oxygen-containing malodorous gases (13.51%).

Table S3 presents the cumulative threshold dilution factor of malodorous gas pollutants in the slag stack. The accumulation threshold dilution factor of malodorous gas in this area is lower than other areas. The main malodorous gas is still sulfur-containing compounds (84.27%), with hydrogen sulfide being the primary component.

Table S4 presents the threshold dilution factor in the flue gas purification area. The total threshold dilution factor in this area is relatively small (2651.52). The main malodorous gases are sulfur-containing malodorous gases, with methyl mercaptan being the primary source (78.86%) rather than hydrogen sulfide.

17.3.4 Health Risk Assessment Analysis of Odor Pollutants

The health risk of mixed environmental toxicants is the sum of the health risks of individual toxicants. According to US Environmental Protection Agency regulations, $LCR < 10^{-6}$ indicates acceptable carcinogenic risk, $10^{-6} \leq LCR \leq 10^{-4}$ indicates potential carcinogenic risk, and $LCR > 10^{-4}$ indicates greater carcinogenic risk. $HI > 1$ indicates non-carcinogenic risk to humans, and $HI \leq 1$ indicates no non-carcinogenic risk to humans.

Table S5 presents the health risk assessment results of the garbage pit. Halogenated hydrocarbons and aromatic hydrocarbons pose the most significant risks to human

Table 17.1 Threshold dilution factor for malodorous pollutants at the feed pit

Category	Compound	Odour thresholds	TDR	Category TDR
NH3	NH3	1.500	27.970	27.970
Saturated hydrocarbon	2-methylpentane	0.420	0.014	0.228
	3-methylpentane	8.900	0.000	
	Propane	1500.000	0.000	
	Pentanea	**	**	
	Butane	1200.000	0.000	
	Heptane	0.670	0.009	
	Octane	1.700	0.041	
	Hexane	1.500	0.075	
	N-heptane	0.670	0.012	
	N-pentane	1.400	0.077	
Unsaturated hydrocarbon	Propylene	13.000	0.011	0.057
	Cyclohexane	2.500	0.046	
Aromatic hydrocarbon	1,2,4-trimethylbenzene	0.120	1.125	45.677
	1,3,5-trimethylbenzene	0.170	0.674	
	4-ethyltoluene	0.008	27.642	
	Benzene	2.700	0.624	
	Styrene	0.035	0.231	
	Toluene	0.330	2.850	
	Meta-p-xylene	0.058	0.232	
	O-Xylene	0.380	0.073	
	Naphthalene	0.009	11.478	
	Ethylbenzene	0.170	0.748	
Sulfur compound	Dimethyl sulphide	0.003	19.084	25,417.19
	Dimethyl disulfide	0.002	40.869	
	Carbon disulfide	0.055	1.023	
	Methyl mercaptan	0.000	1028.027	
	Methyl sulfide	**	**	
	Hydrogen sulfide	0.000	24,328.19	
	Ethyl mercaptan	**	**	
	Ethyl sulfide	**	**	
Oxygenated compound	2-Hexanone	**	**	3277.850
	Acetone	42.000	0.034	
	Ethanol	0.520	66.867	
	Ethyl acetate	0.870	5.088	
	Acetaldehyde	0.002	3205.624	

(continued)

Table 17.1 (continued)

Category	Compound	Odour thresholds	TDR	Category TDR
	Isopropanol	26.000	0.237	
Halogenated hydrocarbon	Chloroform	3.800	0.024	0.427
	Dichlorodifluoromethane	**	**	
	1,1,2-trichloroethane	3.900	0.000	
	1,1-dichloroethane	160.000	0.000	
	1,1-dichloroethylene	**	**	
	1,2,4-trichlorobenzene	**	**	
	1,2-dichloroethylene	**	**	
	1,2-dichloroethane	**	**	
	Cis-1,2-dichloroethylene	**	**	
	1,2-dichlorobenzene	**	**	
	1,4-dichlorobenzene	**	**	
	Dichloromethane	160.000	0.024	
	Chlorobenzene	**	**	
	Ethyl chloride	**	**	
	Trichloroethylene	3.900	0.001	
	Carbon tetrachloride	4.600	0.006	
	Tetrachloroethylene	0.770	0.371	
	Methyl chloride	**	**	
Terpenes	α-pinene	0.018	85.705	444.088
	β-pinene	0.03	42.583	
	Limonene	0.038	315.799	
Total	29,213.488			

health. Although many components of malodorous gas are present in the garbage pit, most are not given by IRIS and are represented by "**" in this article. The HI and LCR values at the feed pit exceeded US Environmental Protection Agency standards for non-carcinogenic risk and lifetime carcinogenic risk hazard index. However, the feed pit is generally a closed system with a small environmental diffusion range, resulting in minimal exposure risk.

Table 17.2 presents the health risk assessment results of odorous pollutants in the leachate pool. Similar to the feed pit, the HI and LCR values of the leachate tank both exceeded US Environmental Protection Agency standards for non-carcinogenic risk and lifetime carcinogenic risk hazard index. However, the leachate pool is also a closed system with minimal environmental spread if managed properly. Negative pressure recovery systems can effectively control health risks in this area.

Table 17.3 shows the health risk assessment results of odorous pollutants in the unloading area. It can be seen from the table the health risk assessment indicators

Table 17.2 Results of a health risk assessment of malodorous pollutants in leachate ponds

Category	Compound	SF	RFD	HI	LCR
NH3	NH3	**	**	**	**
Saturated hydrocarbon	2-methylpentane	**	**	**	**
	3-methylpentane	**	**	**	**
	Heptane	**	**	**	**
	Octane	**	**	**	**
	Hexane	**	**	**	**
	N-heptane	**	**	**	**
	N-pentane	**	**	**	**
Unsaturated hydrocarbon	1-butene	**	**	**	**
	Propylene	**	**	**	**
	Cyclohexane	**	**	**	**
Aromatic hydrocarbon	1,2,3-Trimethylbenzene	**	**	**	**
	1,2,4-trimethylbenzene	**	**	**	**
	1,3,5-trimethylbenzene	**	**	**	**
	4-ethyltoluene	**	**	**	**
	Benzene	0.027	0.040	0.233	9.0E-05
	Styrene	**	**	**	**
	Toluene	**	**	**	**
	Meta-p-xylene	**	**	**	**
	O-xylene	**	**	**	**
	Naphthalene	**	**	**	**
	Ethylbenzene	**	**	**	**
Sulfur compound	Dimethyl disulfide	**	**	**	**
	Carbon disulfide	**	**	**	**
	Methyl mercaptan	**	**	**	**
	Methyl sulfide	**	**	**	**
	Hydrogen sulfide	**	**	**	**
	Ethyl mercaptan	**	**	**	**
	Ethyl sulfide	**	**	**	**
Oxygenated compound	2-butanone	**	**	**	**
	4-methyl-2-pentanone	**	**	**	**
	Acetone	**	**	**	**
	Acrolein	**	**	**	**
	Vinyl acetate	**	**	**	**
	Methyl methacrylate	**	**	**	**
	Tetrahydrofuran	0.007	0.900	0.001	1.5E-06
	Ethanol	**	**	**	**

(continued)

Table 17.2 (continued)

Category	Compound	SF	RFD	HI	LCR
	Ethyl acetate	**	**	**	**
	Acetaldehyde	0.008	**	**	2.4E-04
	Isopropanol	**	**	**	**
Halogenated hydrocarbon	Trichlorofluoromethane	**	**	**	**
	Dichlorodifluoromethane	**	**	**	**
	1,1,2,2-tetrachloroethane	0.20	0.05	0.007	2.6E-05
	1,1,2-trichloroethane	0.056	0.004	0.178	1.4E-05
	1,1-dichloroethane	**	**	**	**
	1,1-dichloroethylene	**	**	**	**
	1,2,4-trichlorobenzene	**	**	**	**
	1,2-dichloropropane	**	**	**	**
	1,2-dichloroethane	0.091	**	**	**
	1,2-dibromoethane	2.00	0.009	1.208	7.8E-03
	1,2-dichlorobenzene	**	**	**	**
	1,3-dichlorobenzene	**	**	**	**
	1,4-dichlorobenzene	0.022	**	**	3.1E-05
	FREON11	**	**	**	**
	FREON12	**	**	**	**
	Benzyl chloride	**	**	**	**
	Dichloromethane	**	**	**	**
	Chlorobenzene	**	**	**	**
	Ethyl chloride	**	**	**	**
	Vinyl chloride	0.03	0.003	0.409	1.3E-05
	Trichloromethane	**	**	**	**
	Trichloroethylene	**	**	**	**
	Carbon tetrachloride	0.053	0.004	9.686	7.3E-04
	Tetrachloroethylene	**	**	**	**
	Monochlorodibromomethane	**	**	**	**
	Methyl chloride	**	**	**	**
Terpenes	α-pinene	**	**	**	**
	β-pinene	**	**	**	**
	Limonene	**	**	**	**
Total				11.72	8.9E-03

HI and LCR of odorous pollutants in the unloading area are lower than those in the leachate pool, but higher than those in the feed pit. It can be seen that the garbage unloading area needs to be strictly controlled due to frequent manual operations. At present, the main plan includes optimizing the airflow organization, keeping the material pit in a negative pressure state, preventing the odor from the material pit from leaking into the unloading hall, and at the same time absorbing the malodorous gas in the unloading hall in time to reduce the concentration in the unloading hall. In addition, the emission of malodorous gas can be reduced by spraying malodorous gas removal liquid.

Table S6 presents the health risk assessment results of odor pollutants in the slag stack. The risk assessment indexes of odorous pollutants in the slag dump are lower than the US Environmental Protection Agency's standards for non-carcinogenic risk and lifetime carcinogenic risk hazard index. The high-temperature incineration and treatment of malodorous gas may have destroyed most of the structures, resulting in a minimal impact on human health risks.

Table S7 shows the health risk assessment results of odor pollutants in the flue gas purification area. The risk assessment index of odor pollutants in the flue gas purification area is significantly lower than the US Environmental Protection Agency's standards for non-carcinogenic risk and lifetime carcinogenic risk hazard index. High-temperature incineration and treatment of malodorous gas may have destroyed most of the structures, resulting in a minimal impact on human health risks. These results demonstrate that high-temperature incineration can significantly reduce the health risk of malodorous gas.

17.4 Conclusion

The concentration and composition of odorous pollutants vary by waste incineration process. The garbage pit, unloading area, and leachate pool are the primary sources of odor emissions, while other areas have relatively low impacts. Sulfur compounds and oxygen-containing compounds are the main types of odorous gases that require control. Health risk assessment indicates that HI and LCR risks are concentrated in the leachate pool, discharge area, and feed pit. Special attention should be given to VC emissions in these areas, and protective measures should be taken to minimize health risks to on-site workers.

Table 17.3 Health risk assessment results of odor pollutants in unloading area

Category	Compound	SF	RFD	HI	LCR	Category HI	Category LCR
NH3	NH3	**	**	**	**	**	**
Saturated hydrocarbon	2-methylpentane	**	**	**	**	**	**
	3-methylpentane	**	**	**	**		
	Propane	**	**	**	**		
	Pentanea	**	**	**	**		
	Butane	**	**	**	**		
	Heptane	**	**	**	**		
	Octane	**	**	**	**		
	Hexane	**	**	**	**		
	N-heptane	**	**	**	**		
	N-pentane	**	**	**	**		
Unsaturated hydrocarbon	Propylene	**	**	**	**	**	**
	Cyclohexane	**	**	**	**		
Aromatic hydrocarbon	1,2,4-trimethylbenzene	**	**	**	**	0.582	2.2E-04
	1,3,5-trimethylbenzene	**	**	**	**		
	4-ethyltoluene	**	**	**	**		
	Benzene	0.027	0.040	0.582	2.2E-04		
	Toluene	**	**	**	**		
	Meta-p-xylene	**	**	**	**		
	O-xylene	**	**	**	**		
	Naphthalene	**	**	**	**		
Sulfur compound	Dimethyl sulfide	**	**	**	**	**	**
	Dimethyl disulfide	**	**	**	**		
	Carbon disulfide	**	**	**	**		
	Methyl mercaptan	**	**	**	**		
	Methyl sulfide	**	**	**	**		
	Hydrogen sulfide	**	**	**	**		
	Ethyl mercaptan	**	**	**	**		
Oxygenated compound	Acetone	**	**	**	**	**	2.7E-04
	Ethanol	**	**	**	**		
	Ethyl acetate	**	**	**	**		
	Acetaldehyde	0.008	**	**	2.7E-04		
	Isopropanol	**	**	**	**		
Halogenated hydrocarbon	Chloroform	0.081	0.010	3.557	1.0E-03	9.17	2.3E-03
	Dichlorodifluoromethane	**	**	**	**		
	Trans-1,2-dichloroethylene	**	0.020	0.296	**		

(continued)

Table 17.3 (continued)

Category	Compound	SF	RFD	HI	LCR	Category HI	Category LCR
	Cis-1,2-dichloroethylene	**	0.020	0.124	**		
	1,1-dichloroethane	**	**	**	**		
	1,2-dichloroethane	0.091	**	**	1.3E-03		
	1,2-dichlorobenzene	**	**	**	**		
	1,3-dichlorobenzene	**	**	**	**		
	1,4-dichlorobenzene	0.022	**	**	4.7E-05		
	Benzyl chloride	**	**	**	**		
	Dichloromethane	**	**	**	**		
	Chlorobenzene	**	**	**	**		
	Tetrachloroethylene	**	0.006	5.199	**		
	Trichloroethylene	**	**	**	**		
	Methyl chloride	**	**	**	**		
Terpenes	α-pinene	**	**	**	**	**	**
	β-pinene	**	**	**	**		
	Limonene	**	**	**	**		
Total				9.759	2.8E-03		

Acknowledgements This work was financially supported by the China Postdoctoral Science Foundation (2023M741516) and the National Natural Science Foundation of China (52266011).

Declaration of Interests The authors declare that they have no known competing financial interests or personal relationships that could have appeared to influence the work reported in this paper.

Appendix 1: Supplementary Material

Supplementary data (Tables S1–S7) associated with this article can be found, in the online version.

References

Cai B, Wang J, Long Y, Li W, Liu J, Ni Z et al (2015) Evaluating the impact of odors from the 1955 landfills in China using a bottom-up approach. J Environ Manage 164:206–214

Crawford JF, Smith PG (2016) Landfill technology. Elsevier, New York

de Titto E, Savino A (2019) Environmental and health risks related to waste incineration. Waste Manage Res 37:976–986

Domingo JL, Marquès M, Mari M, Schuhmacher M (2020) Adverse health effects for populations living near waste incinerators with special attention to hazardous waste incinerators. A review of the scientific literature. *Environmental Research* 187:109631

Fang J-J, Yang N, Cen D-Y, Shao L-M, He P-J (2012) Odor compounds from different sources of landfill: Characterization and source identification. Waste Manage 32:1401–1410

Guo H, Duan Z, Zhao Y, Liu Y, Mustafa MF, Lu W et al (2017) Characteristics of volatile compound emission and odor pollution from municipal solid waste treating/disposal facilities of a city in Eastern China. Environ Sci Pollut Res 24:18383–18391

Hargreaves J, Adl M, Warman P (2008) A review of the use of composted municipal solid waste in agriculture. Agr Ecosyst Environ 123:1–14

Kamaruddin MA, Yusoff MS, Rui LM, Isa AM, Zawawi MH, Alrozi R (2017) An overview of municipal solid waste management and landfill leachate treatment: Malaysia and Asian perspectives. Environ Sci Pollut Res 24:26988–267020

Komilis DP, Ham RK, Park JK (2004) Emission of volatile organic compounds during composting of municipal solid wastes. Water Res 38:1707–1714

Kumar S (2011) Composting of municipal solid waste. Crit Rev Biotechnol 31:112–136

Li X, Zhang C, Li Y, Zhi Q (2016) The Status of Municipal Solid Waste Incineration (MSWI) in China and its Clean Development. Energy Procedia 104:498–503

Lin KL (2006) Feasibility study of using brick made from municipal solid waste incinerator fly ash slag. J Hazard Mater 137:1810–1816

Liu J, Zheng G (2020) Emission of volatile organic compounds from a small-scale municipal solid waste transfer station: Ozone-formation potential and health risk assessment. Waste Manage 106:193–202

Liu Z, Liu Z, Li X (2006) Status and prospect of the application of municipal solid waste incineration in China. Appl Therm Eng 26:1193–1197

Nanda S, Berruti F (2021a) Municipal solid waste management and landfilling technologies: a review. Environ Chem Lett 19:1433–1456

Nanda S, Berruti F (2021b) A technical review of bioenergy and resource recovery from municipal solid waste. J Hazard Mater 403:123970

Nie E, Zheng G, Shao Z, Yang J, Chen T (2018) Emission characteristics and health risk assessment of volatile organic compounds produced during municipal solid waste composting. Waste Manage 79:188–195

Rand T, Haukohl J, Marxen U (2000) Municipal solid waste incineration. A decision Maker's Guide. The International bank for reconstruction and development, World Bank, Washington, DC

Vinti G, Bauza V, Clasen T, Medlicott K, Tudor T, Zurbrügg C et al (2021) Municipal solid waste management and adverse health outcomes: A systematic review. Int J Environ Res Public Health 18:4331

Wu C, Liu J, Liu S, Li W, Yan L, Shu M et al (2018) Assessment of the health risks and odor concentration of volatile compounds from a municipal solid waste landfill in China. Chemosphere 202:1–8

Wu C, Shu M, Liu X, Sang Y, Cai H, Qu C et al (2020) Characterization of the volatile compounds emitted from municipal solid waste and identification of the key volatile pollutants. Waste Manage 103:314–322

Xu P, Chen Y, He S, Chen W, Wu L, Xu D et al (2020) A follow-up study on the characterization and health risk assessment of heavy metals in ambient air particles emitted from a municipal waste incinerator in Zhejiang, China. Chemosphere 246:125777

Zhang H, Schuchardt F, Li G, Yang J, Yang Q (2013) Emission of volatile sulfur compounds during composting of municipal solid waste (MSW). Waste Manage 33:957–963

Zhang Y, Ma Z, Fang Z, Qian Y, Zhong P, Yan J (2020) Review of harmless treatment of municipal solid waste incineration fly ash. Waste Disposal & Sustainable Energy 2:1–25

Zhang Y, Wang L, Chen L, Ma B, Zhang Y, Ni W et al (2021) Treatment of municipal solid waste incineration fly ash: State-of-the-art technologies and future perspectives. J Hazard Mater 411:125132